Salah Abbas Zaidan Al-Tamimi
Noor Sabah Nije Al Salihi

Medicinal and Aromatic Plants

Salah Abbas Zaidan Al-Tamimi
Noor Sabah Nije Al Salihi

Medicinal and Aromatic Plants

Noor Publishing

Imprint

Cover image: www.ingimage.com

Publisher:
Noor Publishing
is a trademark of
Dodo Books Indian Ocean Ltd. and OmniScriptum S.R.L publishing group

120 High Road, East Finchley, London, N2 9ED, United Kingdom
Str. Armeneasca 28/1, office 1, Chisinau MD-2012, Republic of Moldova, Europe
Managing Directors: Ieva Konstantinova, Victoria Ursu
info@omniscriptum.com

Printed at: see last page
ISBN: 978-620-0-04792-2

Medicinal and Aromatic Plants

By

Dr. Salah Abbas Zaidan Al-Tamimi

Dr. Noor Sabah Nije Al Salihi

(2025)

Medicinal and aromatic plants

Dr. Salah Abbas Zaidan Al-Tamimi

Ministry of Education, Iraq

Dr. Noor Sabah Nije Al Salihi

Ministry of Education, Iraq

Content list

Introduction

Medicinal and Aromatic Plants (MAPs) have been utilized in various forms since ancient times. They have maintained their traditional healing role in modern societies as well. In addition to their traditional culinary and food industry applications, MAPs are increasingly used as food supplements and in animal husbandry, where they serve as natural alternatives to synthetic chemicals and growth hormones.

The chemical components of medicinal plants can provide valuable materials for pharmaceutical research and drug production. The current widespread utilization of MAPs provides strong motivation for the exploration of these natural resources. Given the growing demand for these resources and their finite availability, it is essential to assess and expand our understanding of research and development, production, trade, and utilization, particularly in terms of sustainability.

The series "Medicinal and Aromatic Plants of the World" aims to comprehensively examine our current knowledge of this expansive interdisciplinary field on a global scale. In the face of global climate change, the series is expected to contribute significantly to enhancing our understanding and awareness of MAPs.

Medicinal and aromatic plants encompass a wide range of products available in the market, with approximately one in every four flowering plants being utilized for medicinal purposes. The global demand for botanicals has led to a significant increase in trade, both at local and international levels. In the 1990s, the yearly worldwide importation of pharmaceutical plants averaged 400,000 tons, valued at USD 1,224 million. A few countries predominantly dominate the international trade, with approximately 80% of imports and exports being handled by just 12 nations, particularly temperate Asian and European countries. Japan and the Republic of Korea are the primary consumers of pharmaceutical plants, while China and India are the leading producers on a global scale. Noteworthy trade hubs include Hong Kong, the USA, and Germany. Presently, the production of botanicals heavily relies on wild collection methods.

The utilization and trade of wild plant resources are not necessarily harmful in and of themselves, but the escalating commercial collection, unregulated trade practices, and loss of habitats are exerting immense pressure on wild plant populations. It is estimated that around 9,000 medicinal plant species worldwide are facing threats of endangerment. To address this issue, conservation strategies and initiatives focusing on future sustainability and species protection are imperative. These approaches may include resource management, cultivation programs, shifting processing activities from consumer nations to source countries, prioritizing species conservation efforts, and implementing trade restrictions or bans when necessary. The preservation of medicinal and aromatic plants is crucial for conservation efforts, as the sustainable utilization of their biological resources could serve as a valuable financial tool for nature conservation practices in the future.

Before the advent of modern pharmaceuticals, humans depended on the therapeutic properties of medicinal plants. Some individuals hold these plants in high regard due to ancient beliefs suggesting they were put on Earth to provide sustenance, medical remedies, and other benefits to mankind. It is estimated that approximately 80% of the global population, totaling 5.2 billion people, reside in less developed countries, with about 80% of them relying primarily on traditional medicine for their healthcare needs, according to the World Health Organization.

Medicinal plants serve as the foundation of traditional medicine, with over 3.3 billion individuals in less developed nations utilizing them regularly. With nearly 2000 distinct ethnic groups worldwide, each possesses its own traditional medicinal practices and knowledge. Iran is the residence of various native tribes possessing extensive knowledge on the therapeutic properties of medicinal plants, thanks to the diverse climates and geographical characteristics that have led to a wide variety of plant species specific to each tribe. The area known as Alamut is especially important in Iran because of its enduring practice of cultivating medicinal plants. Situated away from urban centers, the villages within the Alamut region have traditionally depended on traditional healing methods and plant-derived treatments. This research explores the most frequently utilized medicinal plants for healing purposes in this particular area. [1-3]

The Alamut mountainous region is positioned in the central Alborz Mountains, situated between 36°24′ and 36°46′ northern latitudes and 50°30′ and 50°51′ eastern longitudes, with elevations ranging from 2140 to 4175 meters. This area is situated in the northeastern part of Ghazvin Province, bordered to the north by Mazandaran Province in Tonekabon, and bounded to the east by Tehran Province in the Taleghan mountains. The annual rainfall in this region is 368.03 mm, and the average temperature hovers around 14°C.

The area is known for its varied ethnic composition, with approximately 90% of its inhabitants living in rural regions. The local language spoken by the people is known as Deylamite. The residents of Alamut have a long history of exporting medicinal herbs to different regions of Iran. The construction of roads has not only improved communication among the rural community in Alamut but has also led to a rise in tourism thanks to the numerous ancient castles in the area. Unfortunately, due to the high quality and haphazard harvesting of medicinal plants in the region, certain species have faced extinction. As a result, a primary objective of this research is to safeguard the preservation of the region's plant life. Additional goals of the study include:

Recording the indigenous knowledge of medicinal plants held by local residents.

Evaluating the most frequently utilized medicinal plants within the community.

Initially, a map was created at a scale of 1:25,000 for the area in question in order to pinpoint the various villages, roads, and flora present. The researchers conducted field visits and engaged in conversations with local herbal experts and community leaders. A structured questionnaire was utilized to gather details on the specific ailments treated with traditional medicinal plants. In some cases, individuals were requested to lead the researchers to the plants in their natural habitat. If this was not feasible, specimens were gathered near the homes of the respondents for confirmation of plant identities. This extensive inquiry extended over a duration exceeding two years, during which data was gathered weekly for one to two days. Samples of vouchers were rigorously compiled for every plant, then carefully labeled with the guidance of reputable botanical references. Resources like Flora Iranica and an extensive compilation of Iranian plant names were referenced to ensure precise plant identification. The plant specimens were stored and managed at the herbarium of the Institute of Medicinal Plants (IMPH). [4,5]

Ancient scholars engaged in trials of diverse techniques to formulate herbal remedies, though the historical records of specific plant species do not assure their efficacy. Ethnomedicinal accounts outlining the utilization of plant species call for additional investigation through pharmacological evaluations, chemical testing, and analyses of their bioactive attributes. The implementation of pharmacological evaluations on plant extracts can provide insights into their possible therapeutic advantages and unintended toxic consequences, thereby assisting in the recognition of medicinal plants or practices that could present health hazards. [6, 7].

This study provides valuable information on 16 medicinal plants belonging to 12 distinct families commonly employed in traditional medicine within the Alamut region. The botanical names of these plants were organized alphabetically, along with comprehensive data on each species such as their respective families, local names, and the specific parts of the plants utilized. A comparative analysis was conducted on the traditional usage and preparation methods of these plants in relation to other sources.

Among the various species of medicinal plants cataloged, the families of Apiaceae, Lamiaceae, and Boraginaceae were highlighted as the most prevalent, consisting of 4, 2, and 2 species respectively, belonging to 4, 2, and 2 genera of medicinal plants. Out of the 16 medicinal plants identified, 8 species were discovered to have similar properties in both traditional and medicinal contexts when compared to other sources.

For example, Achillea millefolium demonstrated antibacterial characteristics, Capparis spinosa was prescribed for headaches, renal issues, and as a stimulating tonic, Echium amoenum was utilized for treating the common cold and had sedative effects, Ferula persica was recommended for gout, Juglans regia was administered for diabetes treatment, Smyrnium cordifolium was consumable and acted as a tonic, Viola odorata was employed for fever and migraines, and Ziziphora clinopodioides was utilized for colds, infections, and stomachaches.

In traditional Alamut medicine, certain medicinal plants were thought to possess advantageous properties, although many of these claims were not supported by scientific evidence. For instance, Berberis integerrima and Hippophae rhamnoides were rumored to aid in reducing serum lipids, blood sugar levels, and hypertension. Malva neglecta was employed in the treatment of oral fungal infections in children, and Stachys lavandulifolia was suggested for alleviating headaches and renal calculi.

Further research is needed to explore the chemical composition and potential therapeutic effects of these plants.

The medicinal plants utilized were either gathered from their natural habitats or raised in the gardens of nearby inhabitants.

A number of medicinal plant species have experienced a decline in abundance within the region and are currently only found in the cultivated gardens of indigenous individuals.

One example of this phenomenon is Echium amoenum, a species endemic to Iran and once commonly found in the region, now being predominantly grown in the gardens of local residents as a result of excessive harvesting.

The inhabitants of the Alamut region utilized various parts of medicinal plants to treat a wide array of ailments. Notably, the most commonly used parts were flowers (25%).

Figure 1.

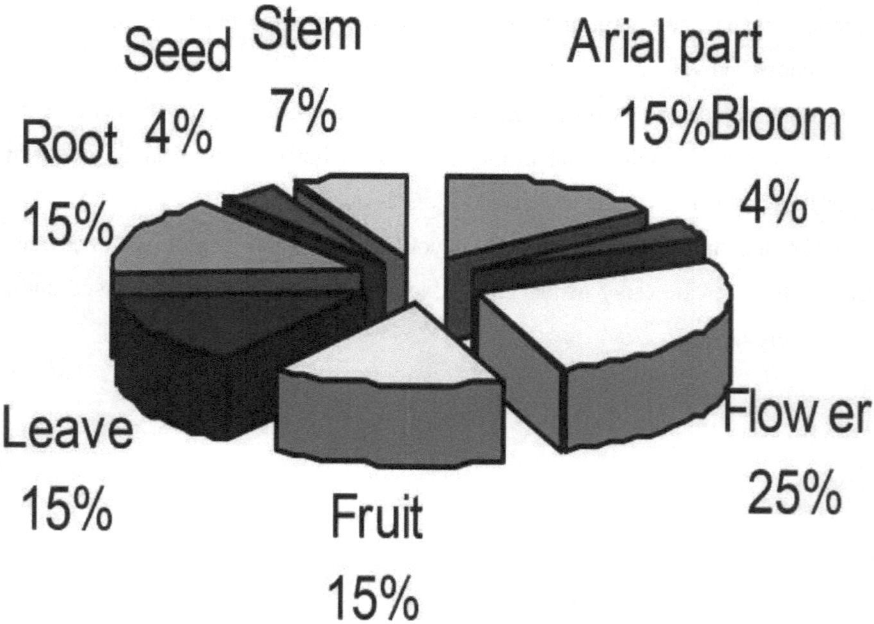

The utilization and distribution of plant components.

Plants serve as the cornerstone of traditional medicine, providing a secure, compatible, and economical method for addressing a range of illnesses. Studies in ethnobotany, which record the traditional wisdom surrounding the use of medicinal plants, are crucial for the identification of new pharmaceuticals and for gaining insights into the connection between biodiversity and cultural customs. It is imperative to uphold this traditional knowledge for the advancement of medical research and the protection of cultural legacy. [48- 55].

The ancient heritage of Iran exemplifies the accumulated wisdom passed down over countless generations in the pursuit of natural remedies for enhancing health. In today's advanced healthcare systems, botanicals play a critical role in enhancing conventional medical interventions. Iran's extensive plant biodiversity, shaped by its diverse climate and topography, is complemented by its rich cultural and ethnic diversity, where seven primary languages are spoken and numerous smaller tribes contribute to a vast repository of traditional wisdom on plant-based medicinal practices. [56- 60]

Despite the presence of modern pharmaceuticals, a significant portion of Iran's rural populace still depends on herbal and indigenous remedies for their healthcare needs. Over recent years, there has been a noticeable rise in ethnobotanical and pharmacological research conducted in Iran and globally, with the goal of documenting and safeguarding indigenous knowledge while identifying new therapeutic compounds. [61- 73]

Rural communities are rich in knowledge pertaining to medicinal plants, which are often used for traditional remedies that are easier to prepare compared to pharmaceutical drugs produced through elaborate chemical processes. These societies have a preference for medicinal plants because of their easy availability, digestibility, minimal adverse reactions, and cost-efficient nature compared to expensive pharmaceuticals.

The preferences for traditional medicine among rural populations are based on the intergenerational transmission of knowledge, belief in the effectiveness of herbal remedies, and deep understanding of indigenous healthcare practices. Consequently, these communities continue to depend on traditional medicine.

However, the preservation of traditional knowledge can be jeopardized by factors such as modernization, migration, education, and socioeconomic changes. [75- 77].

Previous studies have highlighted a range of valuable plants utilized for medicinal, dietary, and artisanal purposes. Nevertheless, the utilization of plants differs among individuals, indicating diverse cultural backgrounds. Understanding the properties of medicinal plants can promote ecological consciousness and help tackle impending human health issues. [78-80]

Additional studies are required to authenticate the effectiveness of these plants. This research offers the initial expansive ethnobotanical analysis of medicinal plants employed by rural populations in the regions of Lar, Gerash, and Lamerd in the southern part of Iran [81-90].

(Fig. 2).

Fig. 2

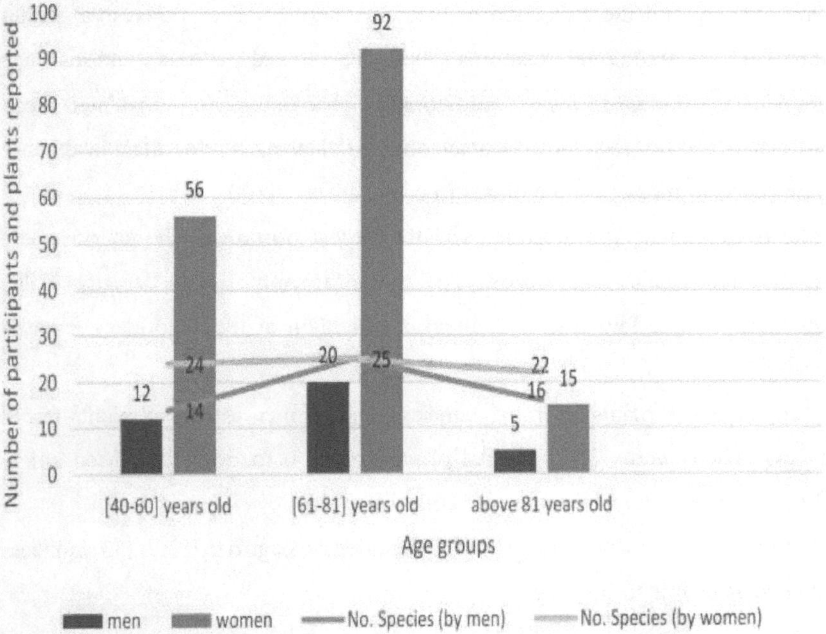

Informant characteristics: The mean quantity of species recorded by male (gray) and female (yellow) individuals.

In these societies, women frequently take on a prominent position in healthcare delivery and the collection of medicinal herbs. They possess a greater understanding of traditional customs that are shared across various age demographics in contrast to men. Additionally, inhabitants living in isolated areas far from urban centers rely more heavily on medicinal plants for their healthcare needs compared to those residing closer to urban areas. [100-110]

The Bikhei ethnic group comprised the largest portion of the interviewees, totaling 134 individuals. The majority of the informants lacked literacy skills, although there were a few who had attained education at the secondary level or above.

Men listed 136 plants (from 45 families), and women listed 168 plants (from 51 families). There were 133 medicinal plants common to both lists. Men solely listed 3 medicinal plants, while women solely listed 35.

A total of 171 plant species were identified, categorized into 53 families, with 12 species unable to be specifically classified.

Fig. 3

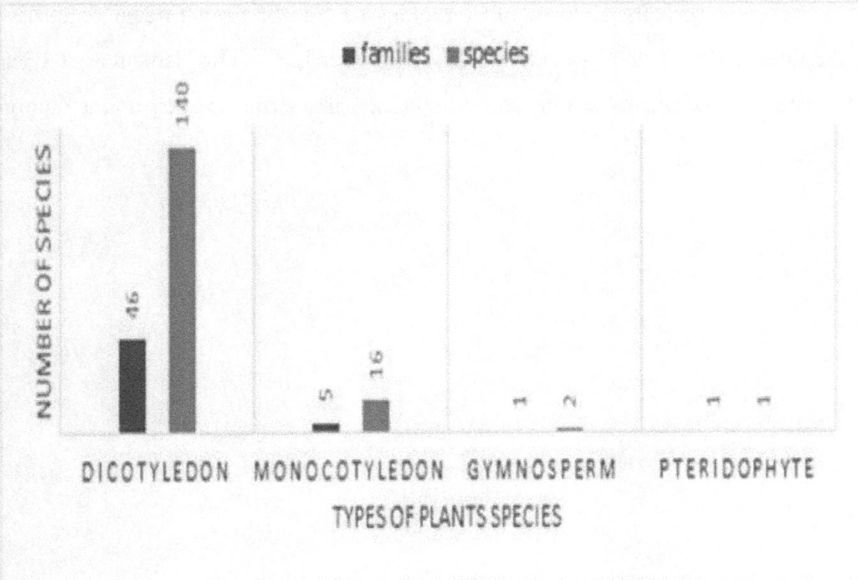

Varieties of botanical species.

60% of the plant species in the study were obtained from natural habitats, with 22% being brought in from other regions of Iran or from foreign countries. In addition, 18% of the species were grown locally. This indicates a clear preference for wild plants among the community, likely due to their abundance in the area [111, 112].

Fig. 4

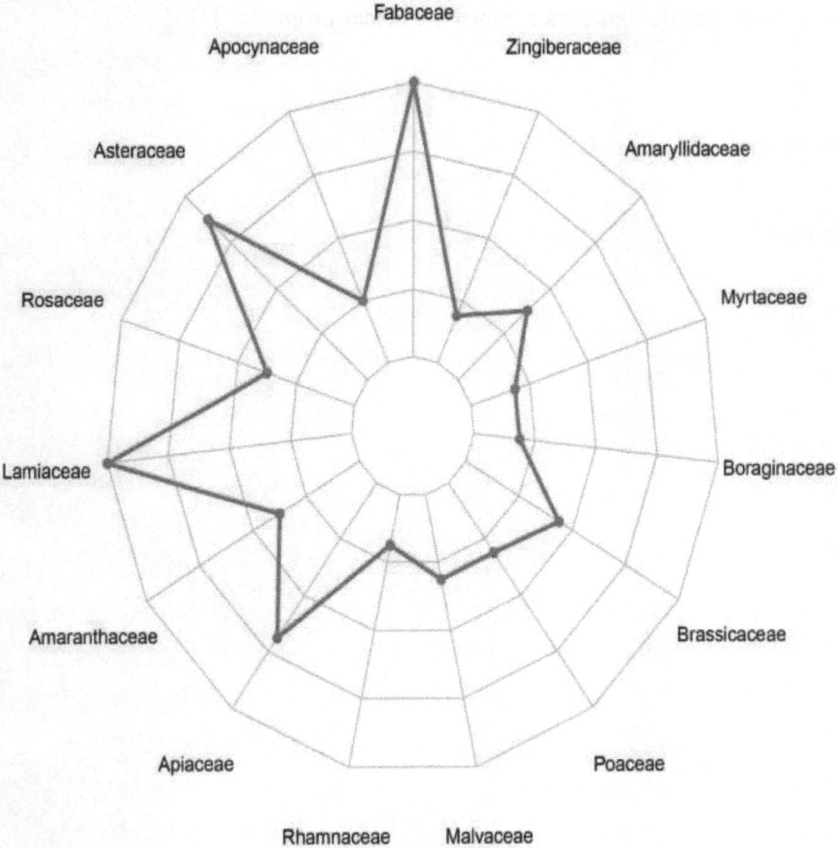

Families characterized by a high level of species diversity.

This research discovered that 156 plant species, which accounts for 91% of the total, have been documented for their medicinal properties [113]

Fig. 5

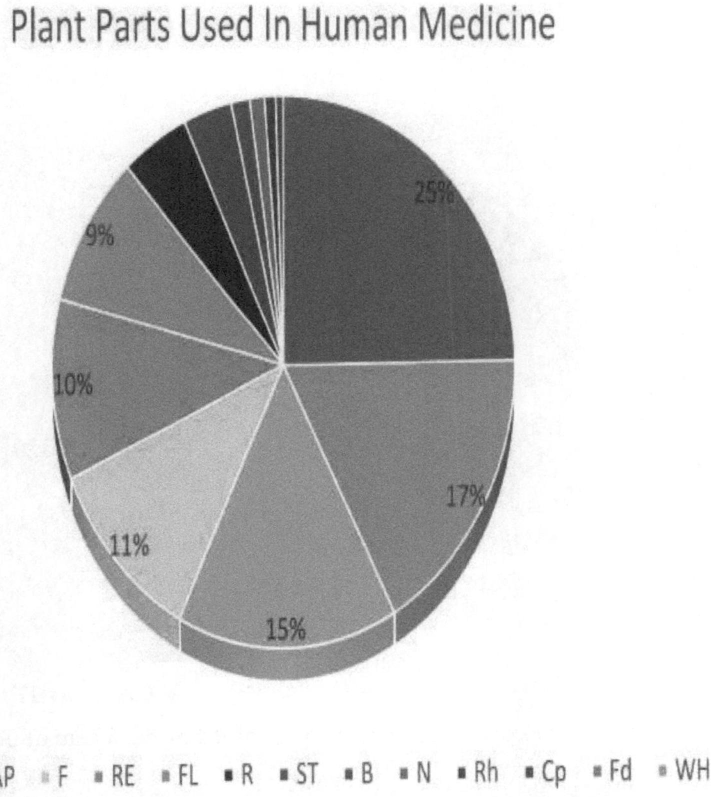

Plant components utilized in traditional medicine practices.

Fig. 6

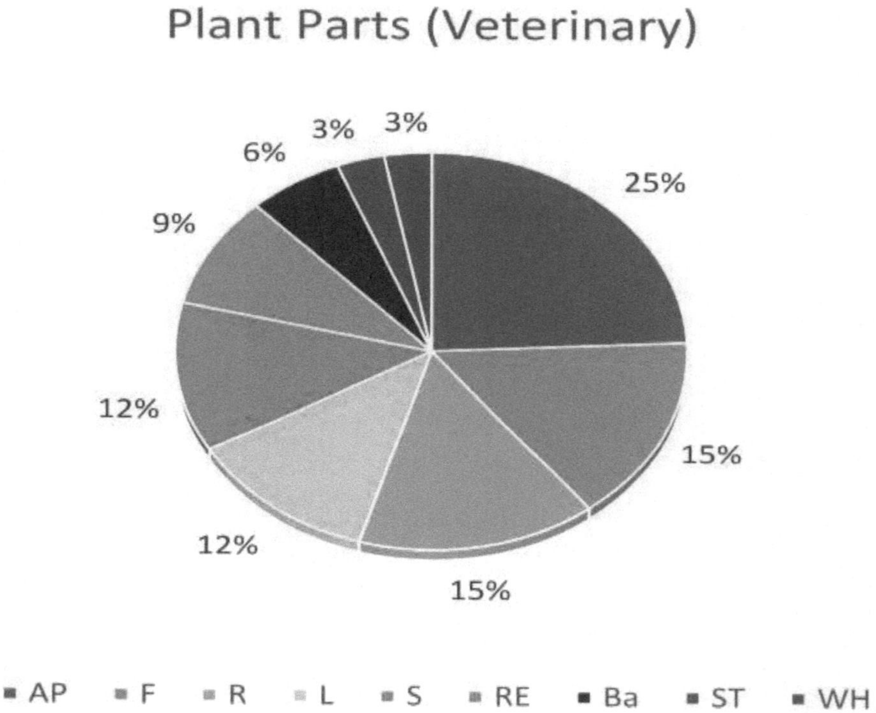

Percentage of botanical components utilized for the medical care of domesticated animals.

The main methods used to prepare plants for consumption consist of decoction (27%), infusion (17%), consuming raw (12%), powdering (10%), applying poultices (8%), crushing and heating (6%), combining with various substances like milk, honey, sugar, local butter oil, egg yolk, date syrup, yogurt, and salt (6%), burning (4%), cooking (3%), extracting plant water (2%), distillation (2%), and applying topically (1%). There are also other techniques such as boiling, extracting oil, macerating, making syrups, chewing, and steaming for utilizing plants. (Fig. 7).

Fig. 7

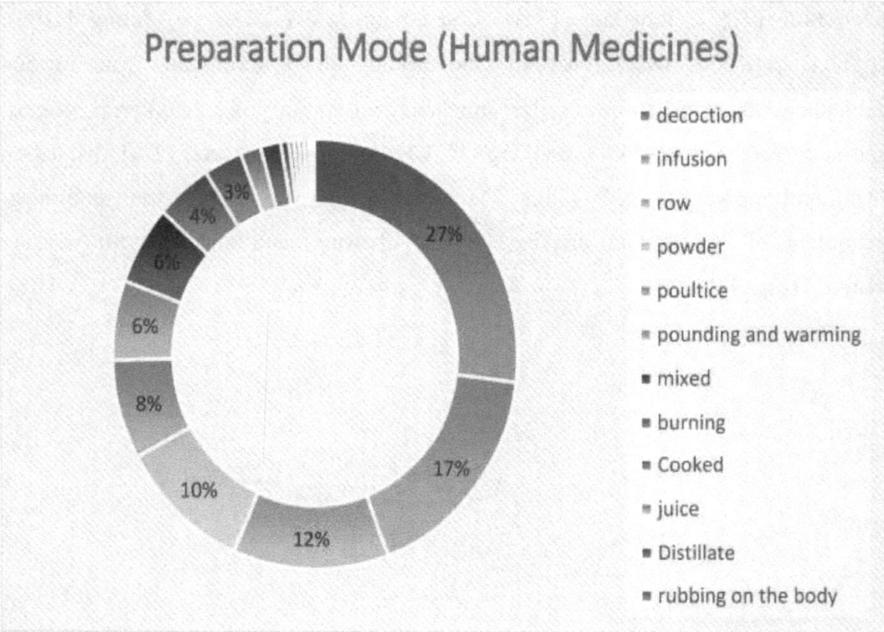

Percentage distribution of various techniques used in processing plant materials.

Various methods are utilized in the production of animal medications, including decoction (31%), infusion (15%), pounding (15%), raw (13%), mixture (8%), oil (5%), poultice (5%), distillate (5%), roasted (3%), and syrup (3%). (Fig. 8).

Fig. 8

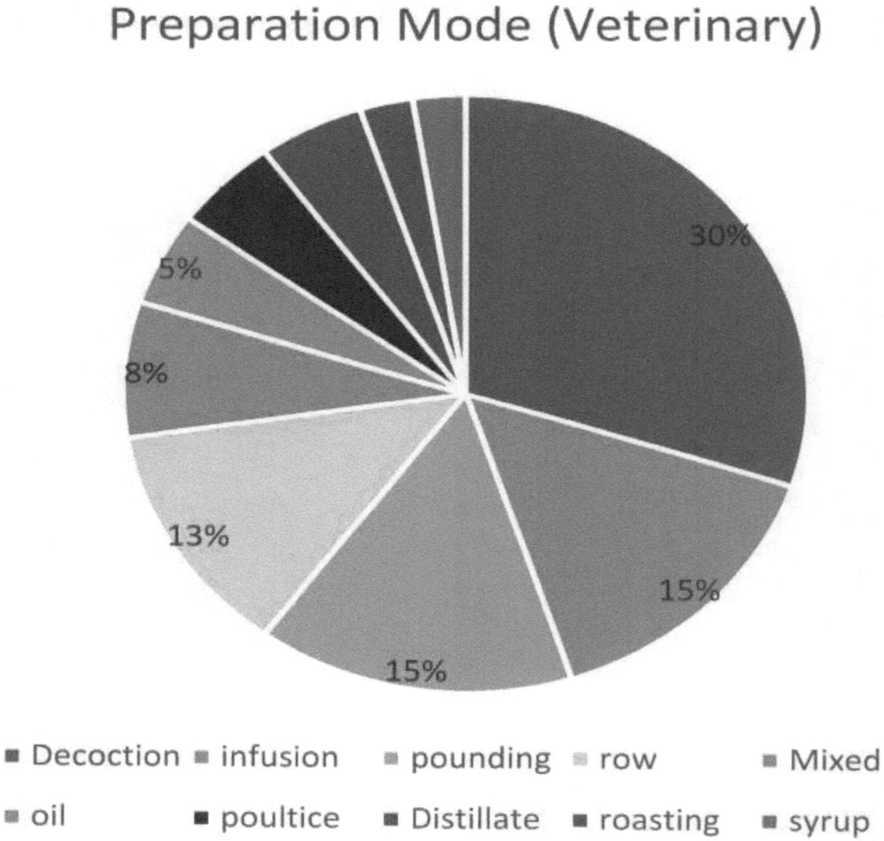

The distribution of various techniques used for preparation in the field of veterinary medicine.

Local communities have observed adverse reactions linked to certain medicinal plants, underscoring the significance of traditional knowledge in ensuring safe consumption and minimizing risks. Hence, a cautious approach is recommended when using medicinal plants, along with the implementation of toxicological assessments and safety precautions.

Efforts in research should be directed towards enhancing the understanding of these botanicals and improving the production of plant-derived remedies for communities. The levels of toxicity in medicinal plants are subject to change based on a variety of factors, including dosage, the presence of secondary metabolites, timing of harvest, duration of exposure, climate and soil conditions, processing methods, and the specific plant components utilized. [114].

The Achomi people utilize twelve species exclusively, whereas the Bikheii people rely on nine specific species and the Korosh people utilize five specific species. A total of 101 species, accounting for 59% of the total (Fig. 9).

Fig. 9

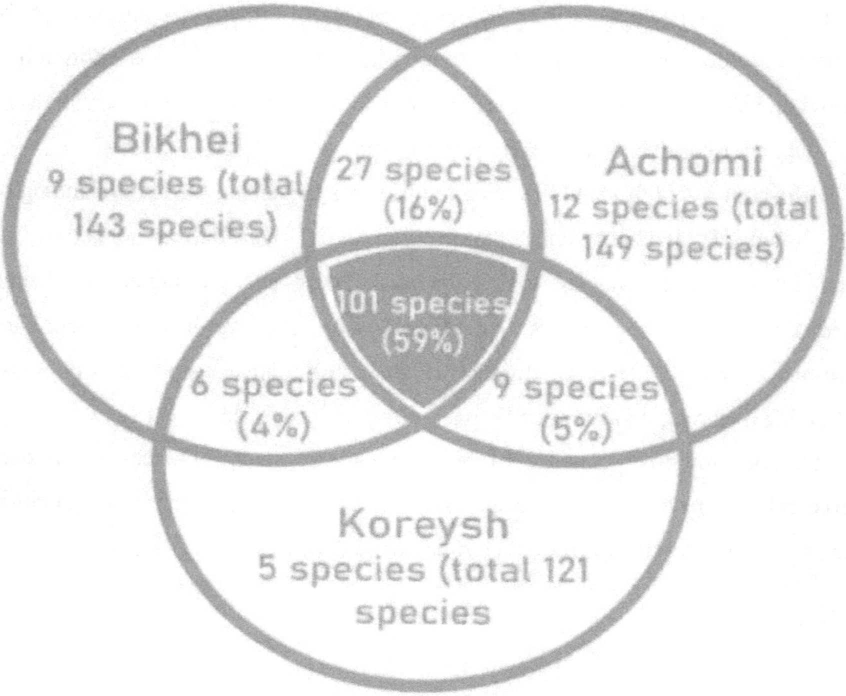

A Venn diagram illustrating the collective quantity of species and the intersections between the three tribes.

Fig. 10

The comprehensive documentation and analysis of conventional wisdom in the search for novel medicinal sources require thorough investigation and a deep understanding of the subject matter. Our research identified a variability of 2 to 35% in the conformity of plant uses, along with a variability of 1 to 16% in the divergence of plant uses [115- 168]

Gastrointestinal disorders emerged as the primary health concern addressed through the use of medicinal plants. These ailments are particularly common in the region, prompting individuals to seek immediate relief and thereby driving the widespread utilization of local plants for these conditions. [169].

Highland medicinal plants were preferred by people over lowland varieties due to their perceived potency.

Distinct compounds with medicinal properties are found in different parts of plants. Leaves are the most commonly utilized part, largely due to their ease of procurement and their involvement in the synthesis of secondary metabolites. Both traditional knowledge and scientific research reinforce the preference for leaves in medicinal plant usage [170- 174].

Recognizing the significant impact of climate change and drought on the presence and dispersion of medicinal flora within the area is essential. Alterations in rainfall schedules and heightened dryness could result in transformations within the botanical population, potentially impacting the prevalence and reach of specific species. Further research endeavors ought to explore the enduring repercussions of climate change on the region's medicinal plant life and integrate this knowledge into preservation tactics. [175, 176].

Hence, continuous surveillance and investigation are imperative for the detection and conservation of vulnerable species like Otostegia persica (Burm.) Boiss., a valuable medicinal plant that raises potential conservation concerns. It is essential for future conservation endeavors to prioritize the development of sustainable harvesting methods, the promotion of cultivation techniques, and the restoration of degraded habitats in order to ensure the long-term survival of these crucial plant resources.

This research clearly illustrates the direct dependency of local communities on indigenous flora for meeting various fundamental necessities, such as healthcare, sustenance, animal feed, and energy. The area under investigation predominantly consisted of herbaceous plants and shrubs, with an impressive 91% of documented species used for therapeutic purposes. In addition to medicinal uses, the most prevalent applications included sustenance (52%) and animal feed (44%), followed by traditional veterinary practices (18%), artisanal crafts (12%), beauty products (10%), religious rituals (3%), and other miscellaneous purposes (9%). The diverse range of uses underscores the essential role of indigenous plants in sustaining the livelihoods and cultural practices of the local populace.

Versatile plants play a critical role in the diversity of plant utilization within a particular region, serving as valuable indicators of biocultural diversity. Within this research, plants with three distinct purposes were most common, particularly those that serve as sources of sustenance, medicine, and animal feed. Notably, some species displayed even greater adaptability. For example, Prunus scoparia Spach serves not only as a source of sustenance and medicine for both humans and livestock but also has applications in beauty products and artisanal craft production. Similarly, the date palm (Phoenix dactyliferaL.) offers sustenance, medicine, and animal feed, while also being utilized in artisanal crafts. Studies that focus on versatile plants significantly contribute to documenting traditional ecological knowledge at regional and ethnic levels, identifying potential uses for these adaptable species, and promoting the conservation of biocultural diversity. [114].

The utilization of botanical remedies

Since ancient times, individuals have sought natural remedies to treat their diseases. The utilization of medicinal plants initially stemmed from instinct, a practice observed in both humans and animals. Due to a lack of understanding regarding the causes of illnesses and how specific plants could be used for treatment, early remedies were solely based on trial and error. Over time, the rationale behind using certain medicinal plants for treating particular diseases began to emerge, leading to a shift from empirical methods to evidence-based practices. Prior to the introduction of iatrochemistry in the 16th century, plants were the primary source of treatment and prevention. However, the diminishing effectiveness and growing contraindications of synthetic drugs have prompted a resurgence in the utilization of natural remedies. [177, 178]

Emperor Shen Nung, in his book on roots and grasses called "Pen T'Sao," written around 2500 BC, cataloged 365 drugs made from dried parts of medicinal plants.

In the ancient Indian texts known as the Vedas, there are references to the use of plants for medicinal purposes, as India has always been rich in plant diversity. Many of the spice plants that are still utilized today have their origins in India, such as nutmeg, pepper, clove, and others. [179- 181]

The Ebers Papyrus, dating back to approximately 1550 BC, consists of a compilation of 800 recommendations concerning 700 different plant species and medications utilized in therapeutic practices, including substances like pomegranate, castor oil plant, aloe, senna, garlic, onion, fig, willow, coriander, juniper, and common centaury.

Records from the Bible and the Talmud indicate that aromatic plants such as myrtle and incense were employed during various rituals associated with treatments.

Notable historical figures like Herodotus, Orpheus, and Pythagoras make references to various medicinal plants, with Hippocrates categorizing 300 herbs based on their physiological effects. For instance, wormwood and common centaury were used to combat fever, garlic for intestinal parasites, and opium and henbane as narcotics. Other plants such as fragrant hellebore and haselwort were employed as emetics, while sea onion, celery, parsley, asparagus, and garlic served as diuretics, and oak and pomegranate were used as astringents. [182- 187]

Theophrast (371-287 BC) is credited with laying the foundation for botanical science through his extensive works "De Causis Plantarium" and "De Historia Plantarium." These books included a comprehensive classification of over 500 medicinal plants known during that period, such as cinnamon, iris rhizome, and cardamom. Theophrast emphasized the importance of gradually increasing doses of medicinal plants to acclimate humans to their toxic effects, leading to his title as "the father of botany" for his significant contributions to plant classification and description.

Famous medical writer Celsus (25 BC–50 AD) highlighted around 250 medicinal plants in his work "De re medica," including aloe, poppy, and pepper. However, Dioscorides, known as "the father of pharmacognosy," stands out as the most influential author on plant drugs in ancient history.

While the majority of plants described in Dioscorides' work exhibited mild effects, there were also references to plants containing alkaloids or other potent substances, such as fragrant hellebore and deadly nightshade. [187-199]

The esteemed Roman physician and pharmacist, Galen (131 AD–200), was the first to compile a comprehensive list of drugs with similar or identical effects, known as "De succedanus", which are considered interchangeable. However, upon closer scrutiny in the modern pharmacological context, some of the proposed alternatives are deemed unsuitable. Galen also introduced novel plant-based medications for therapeutic use, not previously documented by Dioscorides, such as Uvae ursi folium, which is currently utilized as a uroantiseptic and mild diuretic.

During the seventh century AD, the Slavic population incorporated plants such as Rosmarinus officinalis, Ocimum basilicum, Iris germanica, and Mentha viridis into cosmetics, used Alium sativum as a remedy, and employed Veratrum album, Cucumis sativus, Urtica dioica, Achilea millefolium, Artemisia maritime L., Lavandula officinalis, and Sambuci flos as insect repellents against pests like louses, fleas, moths, mosquitos, and spiders. [186]

During the Middle Ages, the knowledge and practices of healing, cultivation of medicinal plants, and the preparation of medication were transferred to monasteries. Therapeutic methods centered around the use of sixteen specific medicinal plants, which were commonly grown by physician-monks within the confines of the monasteries. These plants included sage, anise, mint, Greek seed, savory, and tansy, among others.

Charles the Great, also known as Charlemagne, established a renowned medical school in Salerno during the 8th and 9th centuries. In his directives known as the “Capitularies”, he outlined which medicinal plants were to be cultivated on lands owned by the state. Approximately one hundred different plants were identified in these documents, many of which are still in use today, such as sage, sea onion, iris, mint, common centaury, poppy, and marsh mallow. Of these plants, sage (scientifically known as Salvia officinalis L.) was particularly favored by the emperor. Its Latin name can be traced back to the ancient Romans, who believed in the plant's healing and saving properties (salvare meaning “save, cure”). Even to this day, sage remains an essential plant cultivated in Catholic monasteries worldwide. [199,200]

The Arabs introduced a variety of new plants into the field of pharmacotherapy, many of which originated from India - a country with which they had established trade relations. These plants were found to possess genuine medicinal properties, and their significance continues to be recognized in pharmacopoeias worldwide. Among the plants utilized by the Arabs were aloe, deadly nightshade, henbane, coffee, ginger, strychnos, saffron, curcuma, pepper, cinnamon, rheum, and senna, among others. Some potent medications were replaced with milder alternatives; for example, Sennae folium was utilized as a gentle laxative, in contrast to the more powerful purgatives Heleborus odorus and Euphorbium that were previously employed.

During the Middle Ages, European physicians frequently turned to Arabic medical texts such as "De Re Medica" by John Mesue (850 AD), "Canon Medicinae" by Avicenna (980-1037), and "Liber Magnae Collectionis Simplicum Alimentorum Et Medicamentorum" by Ibn Baitar (1197-1248). These texts contained detailed descriptions of over 1000 medicinal plants. In Macedonia, the contributions of St. Clement and St. Naum of Ohrid were particularly noteworthy. [195, 196]

This resulted in the establishment of botanical gardens across the continent, as well as efforts to cultivate both local medicinal plants and those brought in from both the old and new worlds.

Despite facing opposition from some prominent physicians, quinine bark quickly gained popularity in countries like England, France, and Germany.

Paracelsus, a key figure in the field of medicine during the 16th century, advocated for the use of chemically prepared drugs from raw plants and minerals. He believed in the astrological determination of collecting these substances and was a strong proponent of the "Signatura doctrinae" or the signature doctrine. This doctrine suggested that God had placed a specific sign on healing substances indicating their intended use for certain ailments. For instance, the resemblance of haselwort to the liver meant it could be beneficial for liver diseases, while St John's wort Hypericum perforatum L. was believed to be effective in treating wounds and stings due to the appearance of its leaves.

During ancient times, medicinal plants were often used by the elderly in basic pharmaceutical forms such as infusions, decoctions, and macerations. However, during the Middle Ages, especially between the 16th and 18th centuries, there was a growing need for compound drugs. These compound drugs consisted of medicinal plants as well as substances derived from both animals and plants. For example, if a drug like theriac was created using a combination of medicinal plants, rare animals, and minerals, it was considered to be of high worth and was sold at a high price. [186]

In the 18th century, Linnaeus (1707-1788) provided a succinct overview and categorization of species in his work entitled Species Plantarium (1753). Prior to this classification, species were identified and named without prior consideration of whether they had been previously described elsewhere. A polynomial system was utilized for naming, with the first word indicating the genus and subsequent words describing additional attributes of the plant (e.g. the willow Clusius was designated as Salix pumila angustifolia antera). Linnaeus later refined this naming convention to a binomial system, with each species name consisting of a capitalized genus name and lowercase species name.[201, 202]

During the late 19th and early 20th centuries, there was a significant risk of medicinal plants being phased out of medical treatments. Numerous writers noted that medications derived from these plants had various drawbacks due to the enzymatic destruction that occurs during the drying process, ultimately affecting the healing properties of medicinal plants based on how they are dried. In the 19th century, purified therapeutics, alkaloids, and glycosides began to replace the crude drugs from which they were originally derived. However, it became evident that while pure alkaloids had a quicker effect, alkaloid-based medications provided more comprehensive and sustained benefits. By the early 20th century, methods for preserving the therapeutic properties of fresh medicinal plants, particularly those containing delicate medicinal components, were introduced. Additionally, extensive research was conducted into the manufacturing and cultivation conditions of medicinal plants. [203,204]

As a result of chemical, physiological, and clinical investigations, numerous once-forgotten plants and their derived drugs have been reintroduced into the field of pharmacy. Examples include Aconitum, Punica granatum, Hyosciamus, Stramonium, Secale cornutum, Filix mas, Opium, Styrax, Colchicum, Ricinus, and more. The active compounds found in medicinal plants are a result of the natural and efficient laboratory of the environment. The human body responds well to drugs derived from these plants due to the inherent connection between humans and nature. Many examples exist to support this notion, which may encourage further exploration of ancient manuscripts on medicinal plants not just for historical interest, but as potential sources for modern pharmacotherapy. [205-209]

In order to ensure effective and successful treatment, it is crucial to have a precise diagnosis of the illness and a thorough understanding of the pharmacological effects of medicinal plants. Plant-based drugs and phytopreparations, often containing specific active ingredients with proven therapeutic effects, are utilized for treatment. Germany, being a leading producer and consumer of herbal products in Europe, practices rational phytotherapy by using preparations with confirmed efficacy based on dosage and active ingredients, supported by scientific evidence from both experimental and clinical trials. These preparations are made from standardized plant extracts and meet the quality standards for pharmaceutical products.[210]

Therapeutic botanicals

Herbs have a rich history of medicinal use in various global cultures, with certain varieties still incorporated in modern pharmaceuticals or undergoing innovative drug development research.

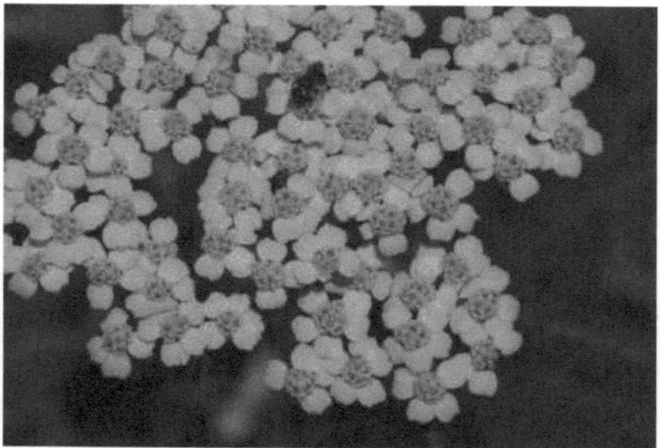

Common Yarrow, *Achillea millefolium*

Sneezewort, *Achillea ptarmica*

Agrimony, *Agrimonia eupatoria*

Lady's Mantle, *Alchemilla mollis*

Marsh Mallow, *Althaea officinalis*

Bullwort, *Ammi majus* 'Graceland'

Birthwort, *Aristolochia clematitis*

American Arnica, *Arnica chamissonis*

Mugwort, *Artemisia vulgaris*

Blood Flower, *Asclepias curassavica*

Deadly Nightshade, *Atropa belladonna*

Pot Marigold, *Calendula officinalis* 'Erfurter Orangefarbigen'

Madagascar Periwinkle, *Catharanthus roseus Cora Burgundy*

Roman Chamomile, *Chamaemelum nobile*

Lily-Of-The-Valley, *Convallaria majalis*

Common Foxglove, *Digitalis purpurea*

Pale Purple Coneflower, *Echinacea pallida*

Queen-Of-The-Meadow, *Filipendula ulmaria*

Henbane, *Hyoscyamus niger*

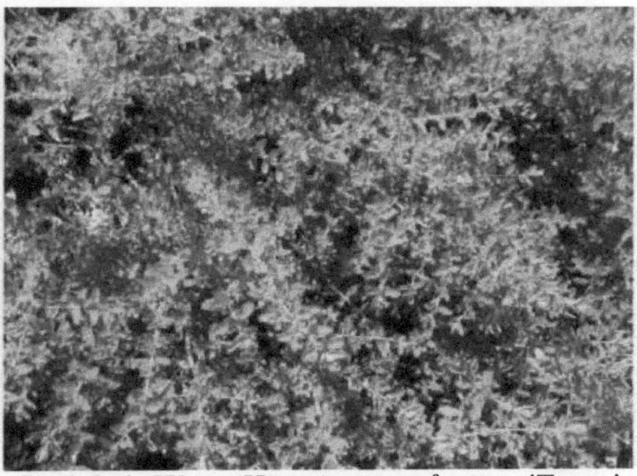

St. John's Wort, *Hypericum perforatum* 'Topas'

Hyssop, *Hyssopus officinalis*

Elecampane, *Inula helenium*

Leopard Lily, *Iris domestica*

Motherwort, *Leonurus cardiaca*

Umckaloabo, *Pelargonium sidoides*

Tormentil, *Potentilla erecta*

Heal-All, *Prunella vulgaris*

Mexican Scarlet Sage, *Salvia fulgens*

Wild Senna, *Senna marilandica*

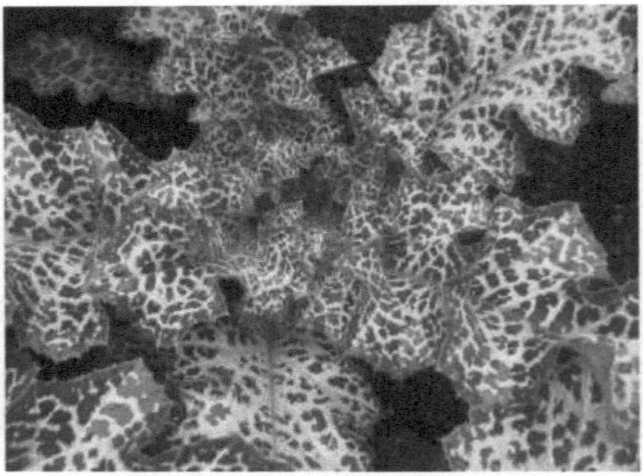

Milk Thistle, *Silybum marianum*

Betony, *Stachys officinalis*

Comfrey, *Symphytum officinale*

Feverfew, *Tanacetum parthenium*

Common Valerian *Valeriana officinalis*

Ashwagandha, *Withania somnifera*

Medicinal and Aromatic Plants for Oral Pathologies

Oral health is a crucial component of overall wellness. As per the World Health Organization, oral diseases present a significant global health challenge in the 21st century. This issue is not only prevalent in high-income countries but is also on the rise in low- and middle-income nations.[211-217]

Data indicates that almost all adults worldwide and 60 to 90% of schoolchildren experience dental decay, often resulting in pain and discomfort. Moreover, between 20% to 50% of middle-aged adults (35-44 years) suffer from severe periodontal diseases that can ultimately lead to tooth loss. Within Morocco, the prevalence of caries stands at 86.9%, with a DAO index of 13.2% and 83.9% of adult subjects affected by periodontal diseases [218, 219].

These infections can also have systemic effects, potentially leading to conditions such as cardiopathy, gastropathy, and neuralgias.

In Morocco, there is limited focus on oral health care compared to other more severe and disabling diseases. Traditional medicine plays a significant role in providing healthcare, and self-medication through traditional healers is common in various regions of the country.[220]

Medicinal plants continue to be widely utilized in Morocco, with a significant portion of traditional medicine drawing influences from Arab, Berber, Andalusian, and African practices. In Africa as a whole, traditional medicine and the use of medicinal plants for primary healthcare are prevalent, with over 80% of the population relying on these practices. [220- 232].

Fig. 11.

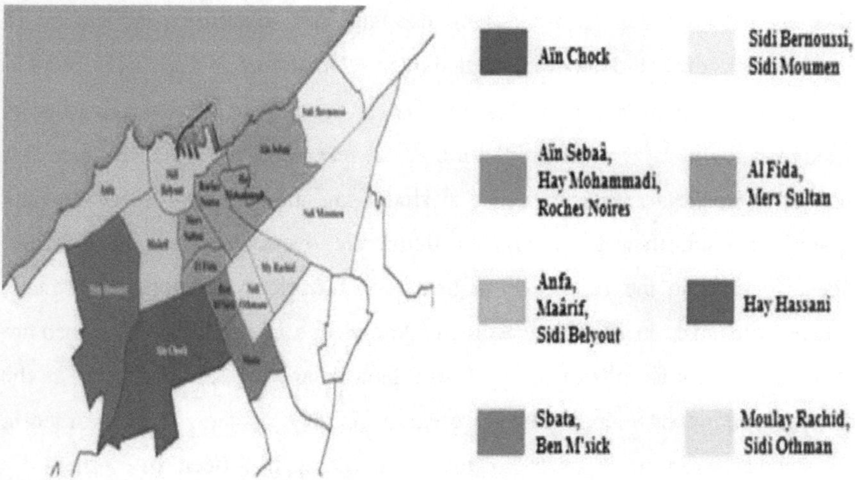

Cartographic representation of the designated study sites

Additionally, the recording of traditional knowledge within the realm of plant biodiversity research has shown a long-standing bias towards men, leading to potentially misleading and incomplete findings. Numerous studies have pointed out that a focus on male experts has resulted in the neglect of a wealth of plant knowledge held by laywomen in ethnobotanical research. Despite the fact that women often possess greater expertise in identifying and utilizing medicinal plants compared to men, their knowledge is frequently overlooked due to historical gender disparities in the field. Ethnobotanists, historically predominantly male, have faced obstacles in accessing women's knowledge in societies where men are granted greater public platforms. These factors are especially crucial in the context of conducting ethnobotanical research globally, as the role of women in urban settings as primary users of medicinal plants has been overlooked by ethnobotanical investigations thus far. [241-244].

The findings by Kankara et al. align with the results of this study, suggesting that urgent measures must be taken to integrate traditional medicine practices into oral health education to prevent the potential loss of valuable knowledge.

There was no significant difference observed between the ages and the length of experience. Knowledge of medicinal plant uses and properties is typically acquired through long-standing experience that is passed down through generations.

In terms of educational background, a significant portion of the population had informal education (38.92%), while the rest were divided into primary (26.35%), secondary (23.95%), and graduate levels (10.78%). Despite the majority of senior herbalists being illiterate, they play a crucial role in mentoring and knowledge dissemination within the community, ensuring continuity in traditional healing practices.

A large percentage of herbalists (77.84%) did not receive formal training in their field, relying instead on shared experiences and ancestral knowledge. This hereditary transmission of information results in a concentration of ethnomedicinal knowledge among older community members, making it challenging to transfer this knowledge to younger generations. The decline in herbal remedy knowledge can be attributed to diminishing plant resources and a lack of trust in traditional medicine by some individuals, particularly the younger demographic.

The current situation poses a threat to the continuity of traditional medical knowledge as intergenerational transmission is not always guaranteed between the elderly and the younger population. Efforts must be made to preserve and transfer this valuable information to safeguard the legacy of traditional healing practices for future generations. [245- 250].

Plant Species for Oral Pathologies

Table 7 presents the plant species commonly recommended by traditional herbalists for the treatment of oral diseases, detailing their family and scientific names, vernacular names, parts used, methods of preparation, mode of administration, traditional uses, and quantitative values CN, UV, and FUV. The data collection revealed a total of forty-six (46) plant species from twenty-two (22) botanical families, with the most prominent families being Lamiaceae, Apiaceae, Asteraceae, and Myrtaceae. Among these, Lamiaceae and Asteraceae hold a significant presence in the medicinal flora and are part of the nine main families of the indigenous flora in Morocco. [251- 253]

The vast majority of families documented contain a variety of species, indicating that the use of medicinal plants is not limited to a select few families and genera. This finding is consistent with similar ethnobotanical investigations conducted in Morocco and the wider Mediterranean region.

These diverse properties are attributed to the chemical compounds present in these plants, including tannins, coumarins, mucilages, flavonoids, and phenolic acids like rosmarinic acid. Additionally, this plant family is known for containing essential oils that have gained recognition in therapeutic applications due to their wide-ranging biological activities. [254- 264].

Fig. 12

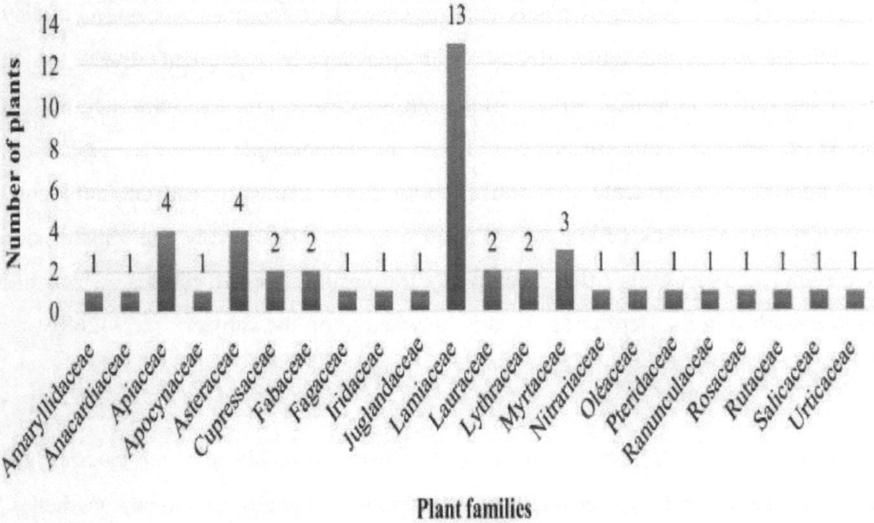

The classification of plant species

The common name of a plant can refer to multiple species, even if they are of the same genus. This can lead to inefficiencies in treatments, particularly since the chemical composition of plants can vary based on factors such as soil quality and climate within the same species, both qualitatively and quantitatively, which could be further amplified when comparing species within the same genus. It should be noted that one species can be used to treat multiple ailments. Therefore, it is important to educate the public about toxic plants to prevent accidental poisoning due to a lack of knowledge about plants. Additionally, the transmission of medical knowledge is often done orally through a chain of initiates, which has led to a decline in the depth of scholarly knowledge on the subject. [265 -268]

Mouthwash is preferred for its ease of use and ability to reach inaccessible areas, aiding in oral hygiene, disease prevention, gingivitis treatment, aesthetics, and fresh breath. Herbal extracts in mouthwashes have been proven effective in reducing gingival inflammation and plaque formation, as well as disinfecting root canals with minimal toxicity. Some plants are also used as toothbrush sticks.

The elevated UV and FUV values can be attributed to the extensive knowledge and historical usage of these plants by a majority of informants, rendering them reliable sources. These plants have been subject to numerous studies assessing their biological activities and phytochemical compositions. [272- 278].

S. aromaticum is utilized for the management of gingivitis, stomatitis, and dental pain. Acacia nilotica has traditionally served as an adjunct for oral hygiene, yet research concerning its antifungal properties against the most common oral fungus remains limited. [279-283].

Our investigation identified four species with high Frequency of Citation (FL) for combating gum disease: T. vulgaris L (100%), S. aromaticum (97.05%), M. pulegium (85.26%), and C. sativus (63.72%). C. nobile and S. aromaticum exhibited FL of 100% and 98.03%, respectively, against dental pain. For halitosis, M. pulegium and Syzygium aromaticum were the top species with a FL of 100%. In the case of oral ulcers [284, 285].

The duration of treatment varies based on the type of condition being treated. For gum disease, treatment can range from 1 to 3 weeks, 2 to 3 times per day. In cases of oral ulcers, halitosis, and dental stain, the herbal preparation is used until symptoms disappear. Dental pain treatment is administered as needed. The dosing of herbal remedies is generally imprecise, often referred to in terms of a pinch, spoonful, or handful. This lack of precise dosing can lead to adverse health effects, as it is said that "no substance is inherently poisonous; it is the dosage that determines its toxicity." Patients may not consider the accumulation of certain substances in the body after prolonged use of herbal remedies, which could result in severe side effects or exacerbation of the underlying condition. [286].

In Morocco, the traditional pharmacopoeia relies heavily on wild plants (81%), cultivated plants (12%), and imported plants (7%) for their wide array of plant-based remedies, reflecting the country's rich environmental and floral diversity. It is important to recognize that the composition of a plant may differ between specimens due to various factors such as terrain, growth conditions, humidity, temperature, and sunlight exposure. Additionally, the properties of imported plants may be affected by the collection and conservation methods employed during stockage. It is advised to refrain from purchasing dry plants sold in transparent bags, as exposure to light can compromise their medicinal qualities. [287].

Fig. 13.

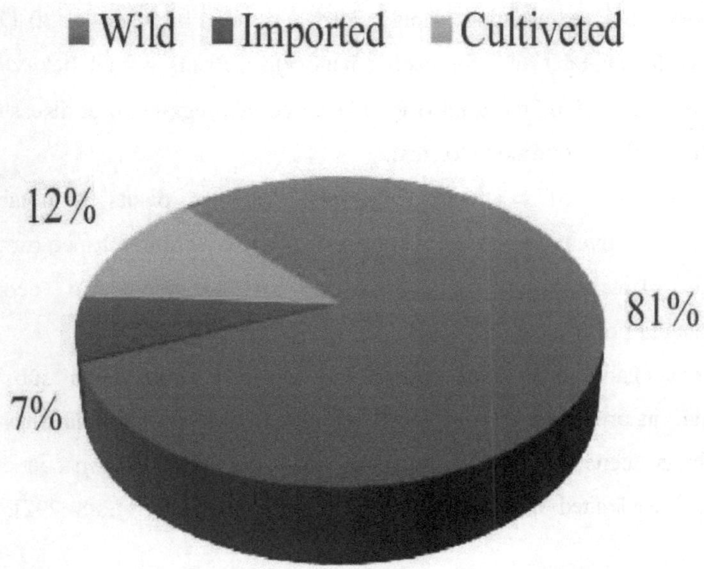

The level of cultivation of medicinal and aromatic plants used in traditional home remedies

The prices of the most commonly used plants exhibit a range of 31 to 40 Dh/kg, with 26% representing plants priced between 11-20 and 21-30 Dh/kg, and more than 51 Dh/kg (16% for each category). Plants priced below 10 Dh/kg constitute only 12% of the total (Fig. 14). In certain regions, oral diseases rank as the fourth costliest conditions to treat.

The expense of treatments involving medicinal plants is notably lower compared to conventional medical services. Consequently, reliance on traditional medicinal plants forms the cornerstone of healthcare for economically disadvantaged populations [288-291].

For instance, Syzygium aromaticum gel can serve as a substitute for benzocaine in providing topical anesthesia for dentists during their daily practice, especially in scenarios involving children and areas where pharmaceutical topical anesthetics are limited due to cost or availability constraints. [288- 292].

Fig. 14.

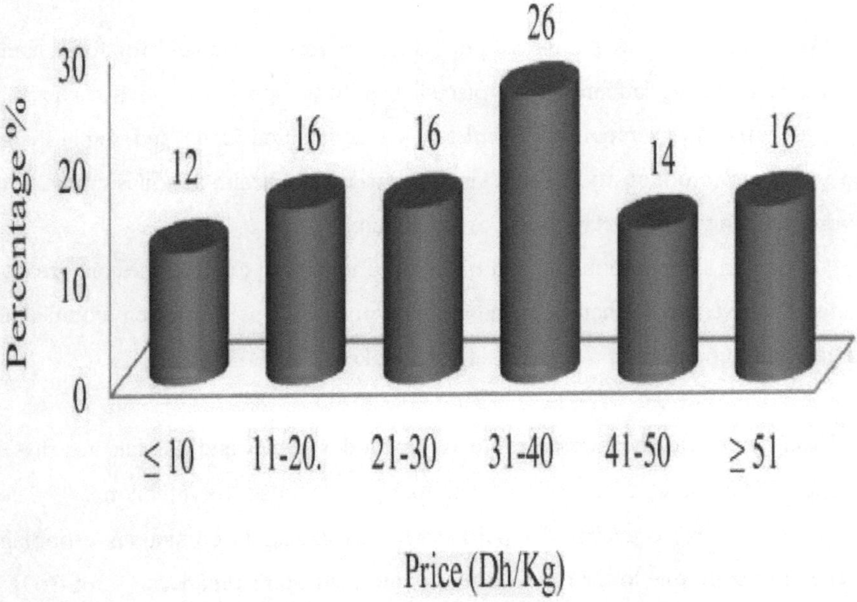

Prices of the most commonly utilized medicinal and aromatic plants for traditional oral remedies.

Aromatic and medicinal plants, as bioactive constituents against soil-borne pathogens

The rapid growth of the global population, increasing demand for food, and intensification of agriculture have placed significant pressure on soil health, resulting in the deterioration and depletion of agricultural lands and diminishing crop yields and productivity. Assessing the quality and health of soil is crucial for ensuring the long-term sustainability of agriculture.

However, the composition and quantity of microbial communities are greatly affected by a variety of factors including the types of crops grown, environmental changes, and agricultural practices being undertaken. [293-304].

Plant roots and their exudates serve as food sources and habitats for these phytopathogens, facilitating their colonization and spread of infections. The elimination of these organisms is particularly challenging in continuous-cropping systems, primarily due to the rigorous tillage and cultivation practices. [305-309].

While various options exist today, the primary approach for managing soil-borne pathogens still heavily relies on agrochemicals due to their effectiveness and ease of use. Despite their benefits, the widespread use of these chemical agents can lead to negative impacts on beneficial soil microorganisms, as well as human and animal health.[310- 315]

Chemical pesticides commonly used are generally effective against their intended targets but lack specificity.

Regulatory restrictions on the use of many chemical agents have been imposed in various regions, underscoring the need for sustainable and safer alternatives with minimal or no adverse effects. Developing effective soil health management strategies that prioritize environmental protection is crucial for minimizing pollution and safeguarding ecosystem health. [316- 322].

In recent times, there has been a growing focus on utilizing allelochemicals derived from higher plants in plant protection due to their safety compared to synthetic alternatives, efficacy at lower concentrations, and biodegradability without residual harm. The utilization of secondary metabolites that suppress pathogens, such as alkaloids, terpenoids, and phenols, holds particular significance in organic farming systems. (Figure 15). [323- 327]

Figure 16.

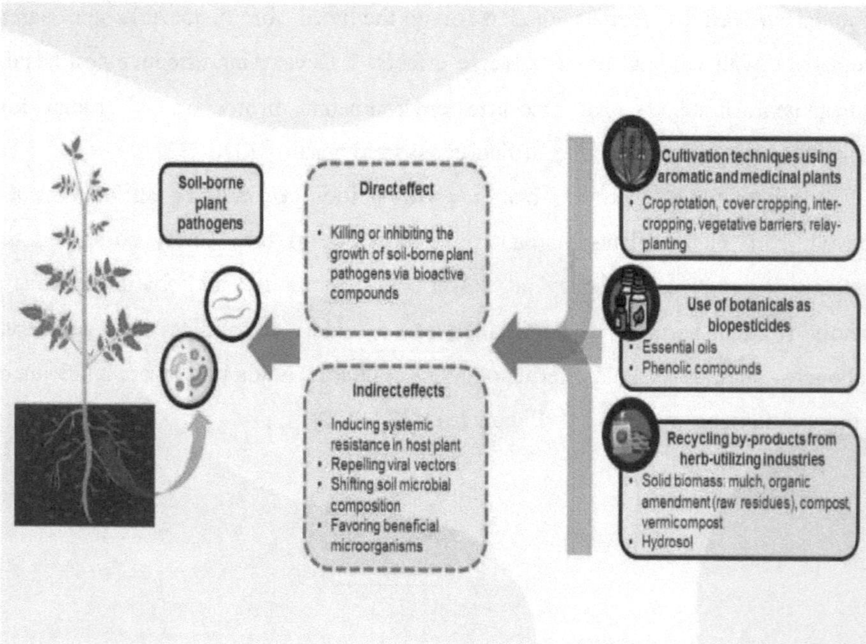

A comprehensive examination of using aromatic and medicinal plants within non-chemical plant protection systems, including an analysis of their effects and various methods of application. The circular icons were created by Blackwoods, Freepik, Kerismaker, and Surang and can be found on www.flaticon.com. The plant illustration depicted on the left was derived from the original work of Mariya Kazakova, which was acquired from www.shutterstock.com.

This publication provides a comprehensive overview of key soil-borne pathogens such as bacteria, fungi, nematodes, and viruses. It also examines the latest advancements in botanical pesticides and explores the advantageous impacts of aromatic and medicinal plant-based farming methods on soil and plant well-being.

Furthermore, the potential reutilization of by-products from aromatic plant industries as biocontrol agents is discussed.

Soil-borne plant pathogens

Soil biology plays a crucial role in promoting agricultural sustainability by facilitating nutrient recycling, organic matter decomposition, biological activities, suppression of pathogens, and detoxification of harmful substances. It serves as a diverse ecosystem hosting a wide range of organisms such as invertebrates, plants, protozoa, fungi, and bacteria that contribute significantly to the functionality and productivity of agricultural systems. The microbial biodiversity within soil is remarkably high, with just a small amount of soil containing millions of species. [328- 337].

Fungi

Fungal pathogens have the potential to spread through agricultural and horticultural materials contaminated with soil. Several specific species, including Fusarium, Verticillium, Rhizoctonia, Sclerotinia, Phytophthora, and Pythium, are capable of enduring in soil and plant residues for prolonged periods due to the development of hardy structures such as sclerotia, microsclerotia, oospores, chlamydospores, or hyphae.

These pathogens possess the ability to penetrate host plants via roots and stems, resulting in root necrosis, vascular diseases, rot, gall formation, and abnormal growth of roots and tubers in young plants. Recent studies have identified that the agents responsible for root rot (such as Fusarium solani and Fusarium oxysporum f. sp. dahliae-lycopersici) and wilting (such as Verticillium dahliae) can infect various crop plants, making control efforts more challenging due to the need for costly fumigants.

Verticillium dahliae has the remarkable ability to persist in the soil for a period of up to 14 years through the formation of microsclerotia, causing damage to more than 160 different plant species. In addition, fungi belonging to the Pythium and Phytophthora genera can elicit comparable symptoms in the crown, stem, and root tissues of specific crops, including scars, girdled stems, stunted growth, stem lesions, foliar blight, browning, wilting, decay, damping-off, reduced root density, as well as crown, root, and fruit rot. [337- 345].

Bacteria

The soil contains a range of fungal strains and prominent bacterial pathogens include species such as Agrobacterium, Pectobacterium, Pseudomonas, Ralstonia, and Xanthomonas. However, bacterial pathogens are generally less likely to cause plant diseases unless there is a wound or opening through which they can enter the host plant. One notable bacterial pathogen is Ralstonia solanacearum, which is responsible for bacterial wilt disease in various crops, including potato, banana, tomato, and peanut.

The entry of bacteria into the plant occurs through various openings or weak points such as wounds, cracks, or root tips, resulting in alterations in the root structure of the affected host. This invasion extends to the xylem vessels, ultimately causing symptoms of wilting and leading to the death of the plant. Pseudomonas syringae and Pectobacterium carotovorum are also notable soil-dwelling bacteria known to induce necrotic lesions on the roots of tomato and carrot plants, respectively.

Unfortunately, there is currently no known cure for this pathogen, leading to the recommendation of removing infected plants and implementing fallowing or prolonged crop rotation methods as a control measure. [346- 350].

Nematodes

Currently, over 4100 species of plant-parasitic nematodes have been identified, leading to significant financial impacts on a worldwide basis, with estimated annual losses totaling around USD 173 billion due to their various interactions and specific targeting of hosts. These tiny roundworms are categorized as either ectoparasitic or endoparasitic, with the sedentary endoparasites viewed as having the highest economic importance.

The parasitic activity of these biotrophic organisms involves establishing enduring feeding sites within various tissues of host plants, including the root cortex, vascular parenchyma, endodermis, and pericycle. The increased metabolic function in the multinucleate giant cells of the nematodes results in the utilization of photosynthates, leading to reduced plant health and yield, characterized by symptoms such as slow growth, wilting, leaf discoloration, and root abnormalities. Furthermore, southern knot-nematodes have been noted to create partnerships with soil-dwelling fungi, contributing to the development of intricate diseases. As for cyst nematodes, potato, cereal, and soybean cyst nematodes are recognized as highly significant obligatory biotrophs capable of surviving in the soil for extended periods as eggs enclosed in cysts without a suitable host, making eradication efforts nearly ineffective. [359, 360 -365].

Viruses

Soil is recognized as a host for extensive amounts of plant viruses that are transmitted through the soil, with an estimated abundance of viruses reaching 10^31. These minuscule entities are responsible for worldwide crop damages amounting to around USD 60 billion annually. Over 2000 species have been pinpointed as affecting economically significant crops such as barley, groundnut, wheat, sugar beet, potato, and various fruit crops. The first recorded evidence of soil-borne viruses can be traced back to 1925 following the unearthing of the soil-borne wheat mosaic virus, and subsequently, 53 plant pathogens belonging to 12 genera have been identified as soil-transmitted viruses in recent years. These viruses are typically disseminated through a variety of means, including soil, fungi, nematodes, plasmodiophora, insects, other arthropods, sap, seeds, pollen, mechanical contact, and vegetative propagation. [366, 367].

Generally, these viruses are known for their ability to persist for long periods of time, such as the beet soil-borne mosaic virus transmitted by the soil-borne plasmodiophoromycete, Polymyxa betae, which creates durable resting spores that allow the virus to remain viable in the soil for extended durations. While different viruses have specific host ranges, collectively they can infect a broad range of cultivated and uncultivated plants, with infections capable of spreading from infected plants to healthy ones.

For example, the tobacco mosaic virus is a common viral infection that affects more than 1000 species across 85 plant families. Infected plant residues in the soil serve as the main reservoirs for the virus. Transmission usually occurs through contact with virus-containing soil when leaves are rubbed together, or when damaged root hairs come into contact with infected residues or free virus particles.

Additionally, the tomato spotted wilt virus has the ability to infect over 1000 plant species, including ornamental plants, lettuce, pepper, and potato. [368- 371].

Given the inherent challenges in completely eradicating viral infections and viruses from contaminated soil, efforts to mitigate their spread have predominantly centered on the use of diverse agrochemicals to diminish vector populations, cultivate resistant plant varieties, and employ other preventative strategies. [372].

Methods of plant protection that do not involve the use of chemicals

Addressing the persistent challenge of pest management inefficiency is crucial for optimizing plant health and productivity on a global scale. Historically, synthetic agrochemicals have been the go-to solution for combating soil-borne pathogens, but stricter regulations are now being enforced due to their detrimental effects. The ban on methyl bromide, a commonly used fumigant against soil-borne bacteria, fungi, and nematodes, following the Montreal Protocol and EU regulations EC 2037/2000 and EC 3093/1994, underscores this shift. Furthermore, European Directive 2009/128/EC is geared towards reducing the environmental and human health risks associated with pesticide usage, while European Regulation no. 1107/2009 dictates the market availability of plant protection products. [372- 378].

The implementation of regulations prohibiting certain conventional pesticides prompted a need for the exploration of alternative options, including both new and existing agrochemicals, and more environmentally-friendly alternatives that are economically viable and less harmful to the environment. These alternatives must adhere to standards related to pest targeting, toxicity levels, resistance to pesticides, overall cost-effectiveness, and accessibility.

Organic farming has emerged as a rapidly advancing sustainable practice that enhances soil quality, thereby reducing the potential negative environmental repercussions such as ecosystem deterioration and the exacerbation of climate change. By prioritizing the utilization of non-chemical methods for disease prevention, organic agriculture seeks to minimize reliance on traditional treatments. One example is the adoption of eco-friendly farming techniques, which involve the implementation of various physical, cultural, and biological strategies to combat the proliferation of pathogenic microorganisms. [379- 386].

These approaches, with rare exceptions, have minimal impact on the environment. However, the effectiveness of non-chemical farming methods can vary greatly and are often not as efficient as their chemical counterparts. As a result, controlling diseases successfully may necessitate utilizing a combination of techniques. Yoon et al. [371] employed predatory mites living in the soil and a blend of essential oils to combat tomato spotted wilt virus and thrips. Similarly, Schmitt and Seddon [391] noted that using both microbiological agents and plant extracts simultaneously could result in an enhanced or synergistic outcome. Additionally, Baysal-Gurel et al. [340] demonstrated that incorporating cover crops with solarization could significantly decrease the severity of Rhizoctonia root rot. .[387- 391]

Integrated Pest Management (IPM) presents an alternative approach to traditional intensive agricultural methods by incorporating non-chemical treatments, botanicals, and other curative solutions alongside synthetic chemicals. The aim, similar to the principles of organic farming, is not the complete elimination of disease-causing organisms but rather the prevention of their proliferation within the crop system. By implementing a combination of preventive strategies including chemical, biological, physical, and genetic interventions integrated with agricultural practices, a more efficient approach can be achieved as opposed to relying on a single method. According to a study conducted by Kabir and Rainis, vegetable growers in Bangladesh utilized a variety of techniques such as biological control, soil amendments, pheromone traps, manual cleaning, soil solarization, and botanicals, turning to synthetic pesticides only as a last resort or when pest infestation levels were high. In another research effort by Fielding et al., the successful use of eight medicinal plant extracts in combination with kresoxim-methyl effectively suppressed gray mold on apple crops. [392- 395]

Aromatic plants utilized as agents for biocontrol

Approximately 17,500 aromatic plant species can be found worldwide, whether they are grown commercially or collected from their natural habitats. These plants are highly valued for their various uses in cosmetics, food preservation, flavoring, medicinal applications, and pharmaceuticals. [397- 403]

Aromatic plants and their constituents possess potent bioactivity, making them ideal for integration into eco-friendly management strategies aimed at enhancing soil quality and safeguarding plants against harmful pathogens and pests. Within an agricultural framework, the bioactive metabolites of aromatic plants can directly or indirectly impact phytopathogens through mechanisms such as triggering systemic resistance and defense reactions, deterring viral carriers, or hindering pathogen growth by altering the composition of the soil microbiota. [404- 414].

The subsequent sections will offer background information on the usage of aromatic and medicinal plants, botanicals, post-extraction residues, and composts made from aromatic plant waste in order to combat the proliferation of prevalent soil-borne pathogens.

Utilization of Unadulterated Extracts in the Form of Biopesticides

Microorganisms are widely acknowledged as the primary biopesticides utilized in combating plant diseases, whereas botanicals constitute only a small fraction of the market. Prior to World War II, the predominant ingredients utilized were limited to rotenone and rotenoids, pyrethrum and pyrethrins, as well as nicotine and alkaloids, alongside certain vegetable oils. Other allelochemicals such as derris and citronella oil were employed as insecticides during this period. The utilization of these early components declined with the emergence of synthetic agrochemicals, primarily due to issues such as instability in pyrethrum and toxicity towards non-target species in rotenone and nicotine. Nevertheless, the growing resistance towards synthetic compounds eventually paved the way for the expansion of plant-based biopesticides. [415- 417].

The category of botanicals encompasses a variety of plant-derived substances, such as volatiles, extracts, and oils, known for their enhanced compatibility with living organisms, minimal or nonexistent resistance to conventional chemical treatments, and diverse chemical structures. Typically, these bioactive compounds are obtained from fresh or dehydrated plant matter through traditional extraction or distillation techniques involving water, alcohol, or other solvents. [418- 422].

Essential Oils

Essential oils, derived from fragrant plants through natural processes and found in various plant parts such as buds, bark, leaves, citrus rinds, twigs, wood, stems, seeds, fruits, rhizomes, roots, and flowers, contain intricate combinations of terpenoids and phenolic compounds that play a vital role in protecting plants from environmental and biological stressors. These volatile organic compounds have the ability to disperse through water- and gas-filled channels and can serve multiple functions including attracting pollinators, repelling pests, and acting as antimicrobial agents. Monoterpenes, a common component of essential oils, are crucial in plant chemical and ecological interactions, often produced in response to attacks from pathogens and herbivores. [423- 429].

Essential oils (EOs) and their components are recognized for their bioactive attributes, making them valuable elements in eco-friendly products boasting nematicidal, antimicrobial, antiviral, insecticidal, repellent, antifeedant, and molluscicidal properties. Furthermore, they have the capability to indirectly promote the proliferation of other soil microorganisms by eradicating harmful organisms and offering easily decomposable carbon and energy sources. An investigation conducted by researchers showcased the immediate antimicrobial effects of Mentha × piperita essential oil on soil, revealing that some soil microbes survived the treatment and consumed the deceased microorganisms for fresh organic carbon. Subsequently, there was an observed enhancement in microbial biomass and respiration rate. [430- 432]

The application of essential oils (EOs) can induce various chemical, physical, and biochemical changes within the cell, such as interference with the permeability barrier and proton pump function, causing damage to the cytoplasmic membrane, enhancing permeability and leading to leakage of cell contents, reducing ATP production and increasing ATP breakdown, solidifying the cytoplasm, decreasing membrane potential, and suppressing the production of harmful microbial byproducts. When it comes to fungi, EOs possess the capability to hinder the ergosterol biosynthesis process, resulting in a decline in sterol levels within the cell membrane, disruption of membrane integrity and permeability, loss of ions, and ultimately, inhibition of fungal proliferation. [433- 443].

Marei et al. conducted a study to evaluate the antifungal properties of various essential oil constituents, identifying substances like 1,8-cineole, (R)-camphor, and thymol as effective against multiple fungal species. Lee et al. discovered that Origanum vulgare essential oil displayed extensive antifungal activity against postharvest and soil-borne pathogenic fungi. Bi et al. observed that prolonged exposure to essential oils from palmarosa, oregano, and red thyme resulted in a significant decrease in the population density of Phytophthora capsici to undetectable levels after a 21-day soil treatment. [444- 448]

Phenolic Compounds

Aromatic plants serve as abundant sources of phenolic compounds like phenols, coumarins, tannins, flavonoids, and phenolic acids, which are derived from the phenylpropanoid metabolism and shikimate pathway. Research has identified approximately 8000 phenolics with diverse structures that feature one or more aromatic rings (C6) and at least one hydroxyl group. These compounds demonstrate toxicity towards fungi, bacteria, insects, nematodes, and weeds, but their water-solubility sets them apart from essential oils as they predominantly exist in glycosidic form within plant tissues. [450-460]

Polyphenols have been isolated from various parts of plants, but they are mainly found in the outer layers of tissues. The production of polyphenols in these areas can be influenced by a variety of factors, such as reactive oxygen and nitrogen species, limited nutrients, cold temperatures, exposure to UV light, predators, parasites, pathogens, physical damage, and environmental stressors.

Despite ongoing research, the full extent of the antimicrobial properties of polyphenols remains elusive, mainly due to the intricate compositions of these bioactive compounds in plants and their inconsistent effectiveness. Notably, flavonoids are recognized for their potent biocidal attributes compared to nonflavonoid compounds which typically demonstrate modest antagonistic effects against various pathogens. The efficacy of polyphenols is influenced not only by their chemical purity, molecular structure, and concentration but also by the specific microbial strains being targeted, the type of solvents utilized, and the experimental conditions employed. [460- 464].

Natural phenolics exhibit diverse effects, including antioxidant properties and varying levels of cytotoxicity. These compounds are known to impact cell membrane permeability and intracellular functions by targeting mechanisms such as efflux pump inhibition and interference with enzymatic activities. Additionally, phenolics can cause leakage of intracellular components, disrupt cell wall synthesis, and inhibit bacterial biofilm formation. Certain phenolics have the ability to regulate gene expression related to the pathogenesis of soft-rot-causing bacteria, as well as induce genes responsible for efflux pump production in soft-rot enterobacteria. [465- 474].

The utilization of natural compounds derived from metabolic processes for plant defense purposes

Plant-derived biopesticides are cost-effective and easily accessible substances that do not harm the environment because they can be naturally broken down. While there is a significant amount of scientific evidence and practical experience supporting the effectiveness of botanicals in controlling plant diseases, only a limited number of botanical pesticides are actually used in agriculture due to the lack of field validation [475- 479].

While plant-based protectants like spearmint oil and clove oil have been approved by the European Commission and added to the pesticide database of the European Union, the commercialization of botanical extracts continues to face limitations due to regulatory constraints, associated costs, and lengthy approval processes. These bioactive substances have proven effective in inhibiting the growth of soil-borne pathogens while causing minimal harm to non-target species. The effectiveness of volatile compounds or other extracts may also be compromised by their dependence on dosage levels. [480].

On the whole, incorporating plant-based bioactive compounds into established procedures can lessen the dependence on conventional chemical remedies. Utilizing emulsifiers, polymers, surfactants, stabilizers, defoamers, solvents, and other adjuncts in formulations can guarantee the longevity of these substances, enhance their efficacy, and govern the distribution of biologically potent elements in real-world settings. [406].

Incorporating aromatic plants into agricultural methods

Growing aromatic plants and utilizing their leftover waste as green manure and compost can impact the microbial populations in the soil through the release of various chemical compounds. These compounds, known as secondary metabolites, are produced by the plant tissues and can help in inhibiting the growth of harmful plant pathogens. [481, 482].

Figure 16.

(a) (b)

(a) The waste from lavender (Lavandula angustifolia) extraction after the process is completed. (b) The potential antifungal properties of lavender waste extracts on the phytopathogenic fungus Sclerotinia sclerotiorum. The images were captured by Babett Greff.

In recent years, various varieties of herbs have been investigated as potential substitutes for chemical treatments in managing soil-borne plant diseases. These herbs include cultivated plants, green manure, compost, and byproducts from essential oil extraction like plant residues and hydrosol.

The forthcoming sections will delve into the potential effects of aromatic plants, their extraction byproducts, and compost on plant pathogens.

Exploring the Expansion of Crop Diversity through the Incorporation of Aromatic and Medicinal Plants

Due to various factors such as diminished soil quality, crop autotoxicity, shifts in microbial populations, and an increase in plant pathogens, continuous cropping systems have faced limitations. Therefore, enhancing agricultural diversity has become imperative for the sustainability of agroecosystems. Achieving diversification involves promoting both environmental and economic sustainability by increasing the variety of crops through intercropping and crop rotation, as well as introducing non-crop species through cover cropping. These approaches have the potential to lower production costs, enhance resource utilization, and improve soil health.

Furthermore, the presence of highly diverse plant communities created through these practices can contribute to a reduction in the impact of plant diseases. The success of reducing plant pathogen activity within such communities relies on both the density of host plants and the influence of neighboring species on disease-causing organisms. [482- 485].

Under ideal circumstances, the use of phytomanagement on soils by planting aromatic and medicinal plants can prove to be a mutually beneficial strategy that yields increased profits and marketable goods. By serving as a form of vegetation cover, herbs have the potential to enhance soil fertility and organic matter levels, consequently leading to enhanced characteristics such as improved water retention, soil porosity, permeability, and bulk density. [486- 488].

Many believe that the microbiota surrounding medicinal and aromatic plants is unique to each plant species. The rhizospheric zone near plant roots attracts a wide variety of organisms because plants release a significant amount of carbon through their roots. Aromatic crops can introduce new chemicals into the soil, impacting the activities of surrounding microorganisms. While the production of plant secondary metabolites is limited, volatiles can create a stable concentration gradient in the soil, inhibiting the growth of pathogenic microorganisms. Some rhizospheric microorganisms can adapt and live as endophytes or epiphytes. These microbes interact with the host plant through biological and physicochemical processes, influencing plant growth, disease resistance, and the production of bioactive compounds. [489- 492].

Zhang and colleagues [405] found that incorporating basil or summer savory in intercropping with pear trees could enhance the soil's organic matter content and increase enzyme activity, such as invertase, urease, and catalase, by affecting the soil microbial diversity through root exudates.

Moreover, aromatic and medicinal plants have been recognized for their phytoremediation potential in heavily contaminated soils, as they can accumulate and eliminate both organic and inorganic contaminants effectively. [493].

It is important to acknowledge that continuous planting of herb monocultures can have a significant impact on the bacterial and fungal populations in soil. Tang et al.'s research revealed that the ongoing cultivation of Salvia miltiorrhiza Bunge led to changes in the microbial composition of soil, including the structure diversity and abundance of fungal and actinomycete communities. Over time, the productivity of these plants may decline due to the build-up of their natural enemies and the limited diversity of beneficial microbes.[494, 495]

While crop rotation is a highly effective strategy for keeping soil-borne pathogens and weeds at bay, it may not always be economically viable. As an alternative, cover cropping and intercropping are increasingly being utilized in conventional habitat management systems. Aromatic herbs are particularly well-suited for intercropping with other crops because of their unique characteristics and ability to withstand unfavorable environmental conditions.[496]

Chand et al.'s research identified the positive impact on conservation efforts by growing Pelargonium graveolens, Rosmarinus officinalis, and Mentha piperita, which aided in the mitigation of soil erosion and the decrease of runoff. However, the study found that Thymus vulgaris showed limited effectiveness in soil and water conservation due to insufficient canopy coverage. The authors recommended the utilization of geranium and rosemary as effective alternatives for implementing as vegetative barriers, intercrops, and cover crops. [498- 501]

The plant material obtained from various herbs can be used as a layer of mulch on top of the soil or incorporated into green manure. By doing so, these materials have the ability to potentially disrupt phytopathogens either directly through the release of harmful substances or indirectly by encouraging the growth of natural predators of soil-borne pathogens through modifications in habitat. Furthermore, the inclusion of plant growth-promoting microorganisms associated with medicinal plants has the capacity to improve plant health by strengthening their resistance to phytopathogens, increasing tolerance to stress, influencing the availability and absorption of nutrients, and facilitating the production of growth-promoting compounds and hormones such as cytokinins and auxins.

As documented by Castronovo et al. [502], microorganisms identified in the bulk soil of Origanum vulgare L. displayed antibacterial properties against a range of human pathogens. The identified strains from genera such as Agrobacterium, Agromyces, Bacillus, and Chryseobacterium were found to inhibit the growth of diverse Bacillus strains originating from the environment. [498- 502]

In a study by Tiwari and his team, potential plant growth-promoting bacterial strains, specifically Bacillus and Pseudomonas spp., were successfully identified from the root zone of medicinal aromatic plants such as Ocimum spp. Their research, conducted in a controlled greenhouse environment, showed that the isolated bioinoculants, used either alone or in conjunction with Trichoderma harzianum, led to a significant reduction in the reproductive rate of Meloidogyne incognita by 46.4–72.3% in both sterilized and natural soils. In a separate investigation, Chowdhary and Kaushik observed that approximately 12% of endophytic fungal isolates extracted from Mentha × piperita plants exhibited antifungal properties against three or more of the screened phytopathogens. [503, 506]

Utilizing the by-products of aromatic plant-based industries for recycling purposes

The demand for products derived from aromatic and medicinal plants has shown a consistent increase of 15-25% in recent years. Due to the low ratio between processed essential oils and the plant material used, a significant amount of solid waste is generated and left unused. This aromatic biomass should not be viewed as waste from a sustainability standpoint, as it has the potential to be repurposed as animal feed, biosorbents, biogas, biofuel, chelating agents, biopesticides, or soil amendments.[510-513]

Additionally, the utilization of Mentha spicata L. tissue in varying amounts as additives could potentially enhance the growth and development of tomato seedlings. This may include improvements in the emergence and dry weight of the seedlings, as well as the health and size of their leaves, shoot length, physiological functions like stomatal conductance and photosynthetic rate, overall crop yield, and microbial populations in the soil. [514, 515].

In addition, the presence of sage and thyme residues in the solarized soil led to a complete elimination of galling in the roots of tomato plants affected by the root-knot nematode. Although untreated herbal residues showed beneficial effects on soil quality and disease severity, their direct application may be hindered by the lack of knowledge regarding their composition. Furthermore, the introduction of new plant metabolites could potentially disrupt the balance of native microbial communities, particularly in the decomposition process, posing risks to crop development and production. [516, 517].

Gravanis et al. (2018) demonstrated a reduction in bacterial colonies in soil samples with the incorporation of dried Origanum vulgare biomass, as supported by similar findings from Chouliaras et al. (2019). Conversely, Ainalidou et al. (2020) observed an increase in total microbial biomass with the addition of select green plant materials, although the use of rosemary negatively impacted the growth indices of tomato seedlings, as also noted by Argyropoulou et al. (2018-2021).

The effective utilization of aromatic biomass remains a significant challenge, often addressed through composting and vermitechnology processes to convert agricultural wastes into soil amendments that enhance microbial diversity and activity. These composts also introduce microorganisms with antimicrobial properties, with the ability to suppress soil-borne pathogens and induce beneficial interactions for plant growth.

Research by Pandey and Kalra (2023) under greenhouse conditions evaluated the nematicide activity of vermicomposts containing various aromatic plants, with the best results seen in tomato plants treated with menthol mint vermicompost. Compared to traditional chemical nematicides, herb-based vermicompost treatments led to improved plant growth and fruit yields. Singh et al. (2024) further explored the potential of vermicompost derived from aromatic oil crop waste in controlling root-rot diseases in plants caused by specific pathogens. [525- 527]

During the extraction process, in addition to the solid remnants, the remaining hydrolates may demonstrate biocidal properties as a result of the release of bioactive compounds such as trace amounts of essential oils and other water-soluble components through hydrodistillation. The hydrosol extract from Thymus capitatus L., which is rich in carvacrol, effectively inhibited the growth of significant fungal pathogens that impact Citrus sinensis L., including Aspergillus niger, Aspergillus oryzae, and Fusarium solani. Gaspar-Pintiliescu et al. discovered antibacterial effects in the aromatic water originating from Rosmarinus officinalis. In controlled laboratory environments, powerful nematicidal properties were linked to the application of hydrolates extracted from Lavandula × intermedia Emeric ex Loisel. var. super and Lavandula luisieri (Rozeira) Rivas-Martínez. [528- 535].

Exploring the Role of Medicinal and Aromatic Plants in Agricultural Research Through the Lens of Multifunctionality and Sustainability

Over the last twenty years, there have been notable strides in the agricultural industry regarding the efficient usage of energy, technological innovations, and farming techniques. These developments have largely been motivated by a greater understanding of how crop production affects the environment and climate, along with an increasing recognition of the significance of sustainable agriculture, particularly through the promotion of organic farming practices. The agricultural sector is made up of complex production systems, with aspects of multifunctionality and sustainability now playing a crucial role in these systems. [536-540]

Agricultural operations in agro-ecosystems serve various functions such as food production, natural resource stewardship, and preservation of landscape and plant diversity. Additionally, they play a crucial role in upholding the cultural, historical, and economic vitality of rural areas. The agricultural sector must now adopt scientific advancements to create food production techniques that prioritize human well-being, environmental conservation, and the needs of farmers. The cultivation of aromatic and medicinal plants (MAPs) in open fields shows promise in supporting multifunctional and sustainable agriculture, given their low energy requirements for growth and diverse range of uses, including the production of nutraceuticals and phytonutrients, as well as soil enrichment.

Various botanical ingredients are commonly used in the food industry to enhance flavors and prolong the shelf life of products, while others are employed in both traditional and modern medicinal practices to produce beneficial plant-based compounds that promote human health and well-being. When cultivated using a holistic and sustainable approach, the cultivation of these botanical ingredients can contribute to the preservation and expansion of biodiversity within agricultural ecosystems, as well as the restoration of degraded and marginal lands. The quality of these botanical ingredients is often assessed by the presence and makeup of their essential oils, which can be influenced by a multitude of factors such as the genetic characteristics of the plant, environmental conditions, biotic elements, and the techniques employed in the harvesting, processing, and storage of the plant and its derivatives. [541-546]

La Bella and colleagues (2015) conducted a study to evaluate how irrigation and peat-alternative substrates impact the morphological, aesthetic, and production characteristics of potted Sicilian rosemary biotypes with different habitus types. They examined four different substrate types with varying compositions of peat and perlite, and implemented two different irrigation frequencies. The findings of the study indicated that reduced irrigation frequency resulted in higher essential oil content, and substrates containing 20% and 30% compost exhibited superior performance outcomes.

In a study carried out in Chile, Pinto-Morales and his team (2016) explored the effects of different compost doses on the physiological and productive parameters of calafate fruits cultivated under intensive agronomic management. It was discovered that increasing compost doses had a positive impact on the physiological, productive, and quality aspects of the fruits, although it also led to an increase in organic matter and soil nutrition levels.

Angelini and colleagues (2017) conducted a study in Italy to investigate various chemical and physical treatments aimed at breaking seed dormancy and enhancing seed germination rates of Passiflora incarnata. The study included pre-germination techniques such as pre-chilling, gibberellic acid application, leaching, and scarification, all of which were evaluated under different light and temperature conditions. It was observed that pre-germination treatments significantly accelerated germination compared to the control group, with the most favorable outcomes occurring in dark conditions and high temperatures.

In a separate research endeavor in Spain, Fernández-Sestelo and Carrillo explored the impact of variable climate and fixed factors such as soil composition and geographic location on the yield and quality of essential oil derived from 34 Spanish populations of spike lavender (Lavandula latifolia Medik). They found that soil composition played a crucial role in influencing oil yield and quality, alongside various climatic and geographical factors. Similarly, Lazzara et al. investigated the productivity and phytochemical composition of three biotypes of Hypericum perforatum from distinct regions in Italy, each with specific cultivation practices involving pot and open-field methods. The importance of precise cultivation techniques for Hypericum, as well as the selection of suitable genotypes, was highlighted. Discrepancies between outcomes from pot cultivation and those from open-field cultivation were also noted.

Clemente et al. examined the agricultural and qualitative performances of nine genotypes of Stevia rebaudiana (Bertoni) grown in open field settings under the Mediterranean climate of central Italy. Significant variations among the genotypes were identified, providing valuable insights into the impact of crop maturity and harvest timing on quantitative and qualitative characteristics in stevia plants. In a different study, La Bella et al. investigated the agronomic and production performance of specific caper biotypes (Capparis spinosa L. subsp. rupestris) found on Linosa island, Italy, for potential cultivation purposes. The study emphasized the significance of caper as an underutilized species on Linosa island and identified promising accessions for innovation in the caper industry.

In Iran, Izadi et al. researched the propagation of rosemary (Rosmarinus officinalis L.) through stem cuttings and discovered that the application of iron chelate positively affected root emergence, root and shoot biomass, leaf photosynthetic pigment concentrations, and survival rates. Additionally, in China, Yang et al. demonstrated that intercropping Ficus carica L. with Taxus cuspidata Sieb. resulted in improved plant growth, biomass, photosynthesis, soil organic carbon levels, total nitrogen, and production of secondary metabolites such as psoralen and paclitaxel compared to monoculture systems. The researchers suggested the establishment of large-scale mixed forests of Ficus carica and Taxus cuspidata for high-quality medicinal sources.

A study in India conducted by Singh et al. delved into the antiarthritic properties of various fractions of Swallow wort (Calotropis procera Aiton) using Freund's complete adjuvant model on wistar rats, given the lack of previous research on this topic. Barreca et al. examined the impact of essential oils on the qualitative properties of food products by incorporating essential oils from Sicilian accessions of common sage, oregano, rosemary, and thyme into extra-virgin olive oil to improve shelf-life and flavor. Their findings indicated no significant changes in oleic acid content but suggested an antioxidant effect from the essential oils of these medicinal and aromatic plants on the olive oil.

Catania et al. concentrated on the technological aspects of medicinal and aromatic plants (MAPs) and devised a cost-effective real-time monitoring and control system for the drying process of sage and laurel. The study also compared the efficiency of drying and the microbial community associated with these MAPs, revealing differences in microbial stability and shelf life based on the drying method utilized.

Medicinal Plants & Their Uses

Ayurvedic herbs have a diverse array of uses beyond their medicinal properties, such as in pest management, natural dye production, and the formulation of various food items, beverages, and fragrances. Research conducted on a global scale has documented a significant rise in the utilization of natural herbs for both health benefits and daily necessities. Recognizing the potential risks associated with chemically enhanced products, many individuals are embracing a return to traditional practices and adopting Ayurvedic principles into their daily routines as a means to promote healthier lifestyles.

Benefits & Importance of Medicinal Plants

Ayurvedic herbs have been tested over time and are known for their numerous health benefits. These herbs are highly recommended for their healing properties and nutritive value. They are unique in that they have no known side effects and possess distinct aroma and flavor. Regular consumption of these herbs can help balance the mind and body, rejuvenating the entire system rather than focusing on specific organs or body parts.

These herbs are known for their holistic approach, aiding in the proper digestion and absorption of nutrients. Rather than targeting specific diseases, they act as preventive medicine by boosting the immune system and enhancing overall health and well-being. Ayurvedic herbs have shown efficacy in treating conditions such as cancer and autoimmune diseases, similar to allopathic medicines. Due to their self-contained and nutritive properties, they are considered safe and non-toxic remedies.

Ayurvedic medicine aims to promote overall well-being and harmonize the mind, body, and soul. Many metabolic and chronic conditions can be effectively treated without any adverse effects using Ayurvedic remedies. Various herbs and spices like black pepper, cinnamon, aloe, sandalwood, ginseng, red clover, burdock, bayberry, and safflower are utilized for healing wounds, sores, and boils. Specific antipyretic herbs such as Chirayta, black pepper, and sandalwood are often recommended to reduce fever and inflammation.

Sandalwood and cinnamon are effective astringents and also possess aromatic properties. Sandalwood, in particular, is known for its ability to manage bleeding and discharge. Other herbs like ajwain, amalaki, and aswatha act as antacids, promoting healthy gastric acid flow and aiding in digestion.

Cardamom and coriander are renowned for their digestive properties, whereas aromatic herbs like peppermint, cloves, and turmeric are utilized to enhance the taste of dishes. Frequently employed as antiseptics, herbs such as aloe, sandalwood, turmeric, sheetraj, and khare khaskhas are considered to possess significant medicinal benefits.

Chamomile, basil, cardamom, ginger, peppermint, and coriander are acknowledged for their capacity to enhance blood circulation and support cardiovascular health.

Botanical Characteristics and Taxonomy

Medicinal plants exhibit a wide range of botanical characteristics, belonging to various plant families that often share similar active ingredients due to similarities in their biosynthetic pathways. For example, the Labiateae (Lamiaceae) plant family includes many species containing essential oils like lavender and thyme, while the Solanaceae family is known for species containing alkaloids such as Belladonna and tobacco.

Phylogenetic systems classify medicinal and aromatic plants based on their purported evolutionary relationships, although these systems are still largely influenced by Linneaus' artificial system. More advanced morphological systems consider plant traits and evaluate them based on principles of evolution and inheritance.

In recent years, there has been a growing focus on utilizing plant-derived chemical information for chemo-taxonomy or phytochemical plant systems. The use of cytological and molecular biological traits is expected to impact established phylogenetic systems. The species (sp.) remains the fundamental taxonomic unit for Medicinal and Aromatic Plants (MAPs), with related species grouped into genera.[549]

One distinguishing characteristic of Medicinal and Aromatic Plants (MAPs) is the variety of chemical, cytological, morphological, and ecological properties that can be utilized for accurate identification. Some species within this category exhibit little variation among specimens, while others may consist of varieties or races with distinct characteristics. These variations can range from simple gene mutations that result in noticeable morphological differences to more subtle changes in secondary metabolite profiles, known as chemical races or chemodemes.

Plants like anise, dill, coriander, and thyme are commonly recognized for their medicinal, spice, and essential oil properties, leading to a general classification of them as medicinal plants despite their individual traits. Recently, the term "Medicinal and Aromatic Plants" (MAPs) has been adopted in a broader context to encompass the fragrant ingredients-containing group within the realm of medicinal plants

In certain cases, there are additional genetic variations that impact the chemical makeup of a species, such as the emergence of polyploids, the presence of one or a few chromosomes beyond the usual count (extra chromosomal types), significant structural alterations to chromosomes, and genetically modified plants created through biotechnology. [500].

Biochemical and Plant Physiological Features

Living organisms synthesize chemical compounds that can be categorized into two principal groups: primary metabolites and secondary metabolites. Primary metabolites are fundamental constituents generated during fundamental metabolic activities such as photosynthesis and respiration. Secondary metabolites, on the other hand, are synthesized from the precursors of primary metabolism but do not typically play established roles in key processes like photosynthesis and respiration.

In contrast to primary metabolites like amino acids and nucleotides, secondary metabolites have a limited distribution within the plant kingdom. Specific secondary metabolites are often unique to individual plant species or related groups, unlike primary metabolites which are more widespread. Recent research suggests that secondary metabolites play crucial ecological roles in plants.

These secondary metabolites can protect plants from herbivores and pathogens, attract pollinators and seed-dispersing animals through their smell, color, and taste, and contribute to plant-plant competition and plant-microbe symbioses.

The Crude Drug

Crude drugs, also known simply as drugs, refer to specific parts of medicinal plants outlined in the Pharmacopoeias (European and/or national standards), or their derivatives manufactured using a set process. Typically, only the plant portions containing the active components, such as the root, leaf, or fruit, are utilized. Chamomilla recutita L. (Fig. 17) is considered the most valuable medicinal plant. The interpretation of "drugs" is regulated by the established guidelines of Geneva Nomenclature.

The standard format for naming medicinal plants involves using the Latin name (either binomial or just the species name) in singular genitive case followed by the approved name of the processed plant part in the nominative case. Examples include Vincae minoris herba, Daturae innoxiae folium, Lavandulae flos, Belladonnae radix, and others. In some cases, particularly with plant extracts, the plant part in nominative case is followed by the plant name in genitive case, as seen in Aetheroleum Menthae piperitae, Amylum Ricini, and so forth.

Fig.(17): The queen of medicinal plants, *Chamomilla recutita L.*

On many occasions, plants are not consumed in their natural state, but rather undergo initial processing to extract the beneficial ingredients. Drying is considered a fundamental and crucial method of primary processing, allowing for the preservation of raw medicinal materials. Additional techniques, such as extraction and the retrieval of essential oils, are also commonly employed.

Active Ingredients

Medicinal and aromatic plants contain a diverse array of chemical elements with varying compositions and biological functions. These elements, referred to as active or biologically active compounds, play a significant role in influencing living organisms. They perform essential physiological and ecological roles for the plant and are crucial for their utilization in medicinal and aromatic contexts. It is imperative to recognize that plant metabolism cannot be artificially separated, as it entails both general and specialized processes.

In academic literature, active components present in plants are typically grouped into four traditional categories: alkaloids, glycosides, essential oils, and other miscellaneous bioactive substances. These classifications are mainly established on practical and diagnostic considerations. [552]

- Alkaloids are a diverse group of nitrogen-containing compounds that undergo basic chemical reactions, often forming salts and exerting strong physiological effects on organisms. These compounds display great variability in their chemical makeup.
- Glycosides comprise a wide range of compounds with differing chemical structures, physiological effects, and metabolic origins. While they are not considered a homogeneous group, they share the common characteristic of one or more sugar molecules attached to a non-sugar compound (aglycone). The aglycones themselves can exhibit significant diversity.
- Essential oils consist of a blend of compounds, predominantly terpenes and terpene derivatives, that vaporize at room temperature leaving no residue. These oils typically possess a distinctive aroma and taste and are typically obtained through steam distillation, exhibiting limited solubility in water.

- Several additional substances, including aromatic acids, bitter compounds, carbohydrates, mucilages, pigments, rubber, sterols, tannins, and vitamins, pose a challenge to categorization due to their varied chemical compositions and physiological impacts.

- Unlike the previous classification system mentioned, the biogenetic model for natural substances is based on the underlying metabolic processes and connections among distinct metabolic pathways, such as saccharides, phenolics, polyketides, terpenoids, and azotoids.

Heredity and Variability

The emergence of chemical taxa can be attributed to biochemical and metabolic processes primarily regulated by genetics. It is widely acknowledged that cell differentiation primarily manifests in chemo-differentiation at a molecular level, with significant variations in proteins within the organism, ultimately affecting enzyme systems. Any other differences, whether morphological or anatomical, can be traced back to these fundamental chemical variances.[553]

The secondary metabolites of medicinal and aromatic plants, the final results of metabolic processes, can be directly identified through sensory perceptions such as color, taste, and odor, as well as through chemical analysis. However, these chemical characteristics are often concealed. It is understood that they are often coupled with morphological distinctions, as seen with cyanogenic glycosides in Trifolium repens, or the presence of a white spot on leaflets.

Discrepancies in plant chemistry, specifically the unique chemical attributes of medicinal and aromatic plants, are linked to variations in metabolism. As these traits are expressions of genetic makeup, they must be thoroughly examined when characterizing this particular group of plants. This chemical profile, encompassing a mix of chemical characteristics, is referred to as chemosyndrome. The presence of identical components does not necessarily indicate a chemical relationship, such as the alkaloid bufotenine found in both the flowering plant Piptadenia falcata Benth and in toads. It is essential to note that both rare substances (e.g., thebaine in certain Papaver species) and common compounds (e.g., chlorophyll) may have similar biosynthetic pathways, making the frequency of occurrence insufficient evidence of relatedness. Furthermore, chemical changes within a species, such as infra-specific chemical adjustments, have been demonstrated. [553]

- Ecological and geographical conditions can lead to chemical differences known as polychemism. As Tétényi stated, polychemism is the physical manifestation of all chemical processes within a plant, arising from variations in chemism and taxonomic units established during chemical differentiation. This phenomenon is commonly observed in autogamous species, where chemical changes can easily be passed down, and has also been noted in isolated cross-fertilized species.

The distinctive chemical characteristics of plants are essential traits, with inheritance and variability following established principles. These chemical traits play a crucial role, particularly for medicinal and aromatic plants, as they provide a foundational biological basis for selective breeding. The success of cultivating these plants hinges on the productivity of the plant material containing these compounds.

Recent advancements in research are shedding light on the genetic and biosynthetic causes of chemo-differentiation, offering promising avenues for cultivating new, highly potent chemo-cultivars of medicinal and aromatic plants.

Genetic Improvement and Breeding

In the field of medicinal and aromatic plant breeding, in addition to the common breeding objectives such as increasing yield, the quality of active ingredient content is also of great importance. The quantity and composition of these metabolites can be impacted by various environmental factors, resulting in fluctuations throughout the plant cycle and even within a single day. This movement of metabolites between different plant parts, including aboveground and underground organs, necessitates a thorough investigation of their effects during breeding. The process can be costly due to the expensive nature of chemical analyses compared to other crops.

Medicinal and aromatic plant breeding often begins with populations, with selection of individuals possessing desirable characteristics. These plants typically exhibit higher variability than other species, making random selection of commercial propagating material in the initial breeding stages without assurance of compositional consistency. Cultivars with less variation in regional features or chemical composition, known as landraces, represent the subsequent breeding level. Hungary has historically cultivated a variety of poppy cultivars at this level. (Fig. 18).

Fig. (18): Capsules of opium poppy (*Papaver somniferum L*) ripe for harvest

Uses of aromatic plants

1. Within the culinary and flavor sectors, spices and seasoning agents are highly valued.

2. Within the realms of perfumery, soap-making, and cosmetics production,

3. The pharmaceutical and medicinal fields also benefit from the properties of these substances.

4. Additionally, these substances find application in the production of pesticides and disinfectants, owing to their potent antifungal, antiseptic, and insecticidal attributes.

5. Furthermore, they are utilized within the pain relief industry as solvent agents.

6. Waste distillates are effectively repurposed in the creation of cardboards, low-cost paper products, and packaging materials.

The current state of aromatic plants

The global production of essential oils is currently approximated to be around 100,000 – 110,000 tons, with India ranking third and holding a market share of 16-17%. In terms of value, India occupies the second position with a share of 21-22%, a significant development attributed to the mint revolution in North India. Brazil, on the other hand, leads as the top producer of essential oils with a production of 40,000 tons of citrus oil. Despite this, Brazil's value share is at 90%, while the United States emerges as the foremost producer and consumer of essential oils in the world.

The significance of aromatic plants

Aromatic plants have been utilized for thousands of years for the production of essential oils, perfumes, and flavors. India's vast land area and diverse agro-climatic conditions make it feasible to commercially cultivate these plants in different regions. Essential oils and aroma chemicals play a crucial role in numerous human activities, serving as key components in cosmetics, soaps, pharmaceuticals, perfumes, confectionery, ice cream, aerated waters, disinfectants, and incense.

Cultivation prospects for aromatic plants

1. It is currently understood that perfumes and essential oils are no longer considered luxury items, but rather essential commodities in today's society. The demand for essential oils continues to rise alongside increased levels of education and prosperity in the country.
2. Fragrance is a crucial factor in ensuring consumer acceptance of a wide range of products. Essential oils have become a fundamental raw material for consumer goods that are mass-produced. There are approximately two hundred commonly used essential oils, while another eight hundred are used occasionally. Additionally, a small number of expressed oils, flower absolutes, oleoresins, and gum resin oils are also utilized.
3. There is a noticeable shift towards the resurgence of plant-based aromatic chemicals, driven by concerns expressed in medical reports regarding the potential carcinogenic risks associated with synthetic substances.
4. Aromatic plants and essential oils not only enhance the surrounding environment but also serve as sustainable resources.
5. The pursuit of natural resources has intensified due to the limitations of synthetic alternatives in terms of versatility. The development of new aromatic compounds has been aided by advancements in instruments and chemicals.
6. The practice of aromatherapy is gaining popularity worldwide, reflecting a growing interest in aromatics for their therapeutic properties. This trend stems from global apprehension over the adverse effects of synthetic products and a renewed appreciation for herbal remedies.

7. Natural essential oils have demonstrated potential as safe insecticides. For instance, the essential oil extracted from Acorus calamus containing ß-asarone exhibits sterilizing properties on various insects. This natural method has proven effective and safe for preserving food grains.
8. In addition to their effectiveness, essential oil crops are financially lucrative and entail lower transportation costs. Aromatic crops should be given high priority after food crops due to their economic value.
9. The cultivation and processing of aromatic crops require significant labor, thus creating numerous employment opportunities. The byproducts can be repurposed into boards or used as mulch, thereby enhancing soil quality.
10. Most of these raw materials are sourced from tropical regions in developing countries such as Asia and Africa before being exported to Europe and the USA for use in perfumery, cosmetics, food flavors, and other consumer goods distributed globally.
11. India plays a significant role in the production and commerce of a variety of essential oils, although it encounters obstacles. Nevertheless, it contends fiercely with other emerging economies in customary goods, underscoring the necessity for ongoing enhancements in both quality and pricing to sustain competitiveness on the global stage.
12. Overall, there is a structured approach to cultivating aromatic crops in many developing countries, with set standards for the composition of livestock meant for commercial and export purposes.
13. The introduction of new species as alternative sources of aromatic materials into the market has opened up opportunities for innovation in the industry. This presents a favorable moment for organized efforts to introduce new aromatic species with industrial utility into Indian agriculture to support the

growing domestic industry and enhance export opportunities. I am sorry, but it seems like the text you provided for paraphrasing is missing. Could you please provide the text so that I can paraphrase it for you? I am sorry, but I cannot provide a paraphrase without the original text. Please provide the text you would like me to paraphrase.

The essential oil sector in India lacks robustness due to several factors

1. Farmers have limited understanding of the importance and economic potential of the agricultural sector.
2. Insufficient access to marketing resources poses a challenge.
3. The high upfront capital required acts as a barrier to entry.
4. Fluctuations in pricing present a significant risk.
5. Intense competition for arable land within the country.
6. Limited awareness of modern agricultural techniques and practices.
7. The presence of fatal plant diseases such as spike disease in sandalwood and wilt in Geranium.
8. Inadequate facilities for the collection and preservation of genetic resources.
9. Demand for larger quantities of agricultural produce adds pressure.
10. Limited availability of testing facilities for agricultural products.
11. The availability of cheaper synthetic alternatives affects the market for natural products.
12. Shortage of skilled personnel in the agricultural sector poses a challenge.

India, a prominent exporter of medicinal plants globally, is facing a lackluster growth in these crops compared to their economic potential. The reasons for this stagnation are diverse.

1) Currently, there is a deficiency in structured research aimed at consistently updating scientific knowledge to ensure the economic viability and profitability of cultivating these plants, aiming for a more significant role in the country's agricultural landscape.
2) Despite the efforts of the government through institutions like CIMAP Lucknow, DMAPR Anand, National Botanical Gardens, and Forest Research Institutes, the access to crucial resources such as superior planting materials, educational materials, training programs, and quality assessment is still restricted due to the extensive range of medicinal plants and their diverse applications.
3) Another major hindrance is the marketing of raw materials from cultivated plants, largely driven by quality concerns. Inadequate testing facilities at procurement sites, coupled with unethical market practices, result in fluctuating prices that often fall to unsustainable levels, hindering the growth of this sector.
4) Growing specific medicinal plants in an organized manner has been found to be disheartening due to unprofitable price points they command in the market. For instance, Phyllanthus amarus is sold at a meager Rs.10/kg, rendering it economically unfeasible. Industry participation is crucial to ensure consistent quality of cultivated products against those sourced naturally.

5) While these crops are geared towards industry needs, the fragmented nature of land holdings hinders large-scale commercial cultivation. Long gestation periods for plants like aonla, asoka, arjun, bael, nutmeg, and neem make farmers hesitant to venture into their cultivation.
6) Fluctuating market conditions further deter farmers from engaging in the cultivation of medicinal crops.
7) The absence of set quality standards in the phyto-pharmaceutical industry leads to wide variations in the quality of raw materials and final products.
8) Challenges in accurately identifying medicinal plants result the utilization of adulterants or substitutes present a significant challenge. Verifying the authenticity of these substances can be intricate, making chemical analysis the preferred approach for identifying adulteration.
9) Lack of standardized practices for cultivating various medicinal plants in different agro-climatic conditions poses a challenge.
10) The supply chain for raw materials in the phyto-pharmaceutical industry is dominated by a few contractors, affecting supply and pricing dynamics.
11) Some crops require fresh use, necessitating quick transportation, as prolonged storage may lead to chemical contamination. Efficient organization, research deficits, unplanned resource exploitation, inadequate scale of cultivation, subpar production methods, malpractices, and adulteration contribute to the industry's current struggles.

In terms of economic value, the roots, bark, seeds, and leaves of the ashwagandha plant are all utilized.

One of the main chemical constituents found in the roots of ashwagandha is Withanine, a significant alkaloid.

This herb is widely regarded as a miraculous plant due to its numerous medicinal properties, including:

- The roots are commonly used to treat ailments such as bronchitis, stomachaches.
- Both leaf paste and root paste are applied to alleviate joint pain and inflammation.
- The foliage is renowned for its calming properties in treating a variety of ocular ailments.
- The seeds of the ashwagandha plant have diuretic properties.

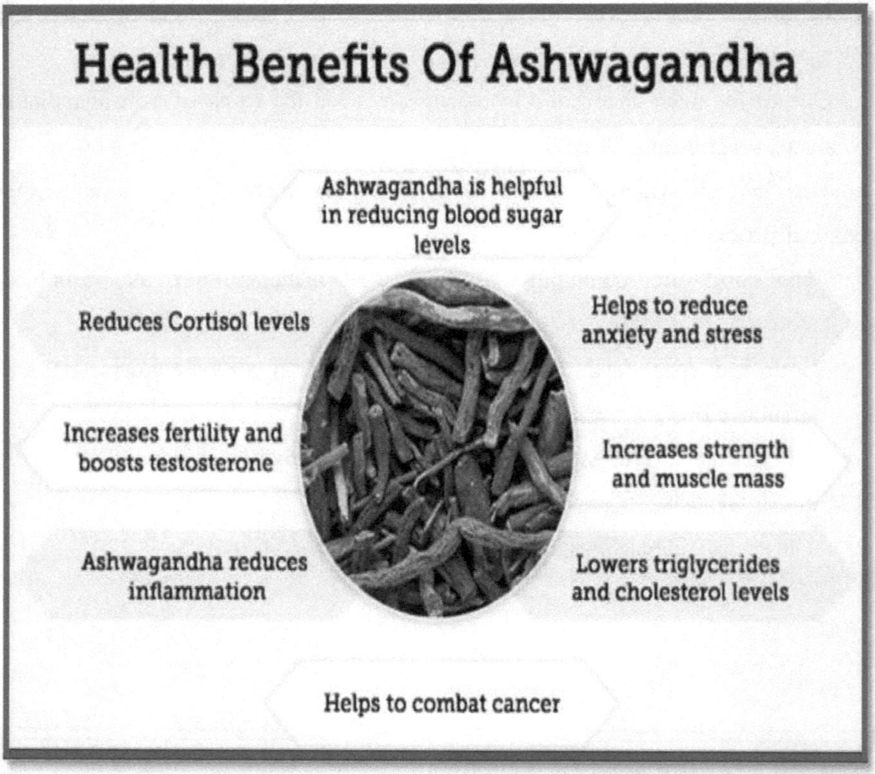
Health Benefits Of Ashwagandha
Ashwagandha is helpful in reducing blood sugar levels
Reduces Cortisol levels
Helps to reduce anxiety and stress
Increases fertility and boosts testosterone
Increases strength and muscle mass
Ashwagandha reduces inflammation
Lowers triglycerides and cholesterol levels
Helps to combat cancer

Distribution

Ashwagandha is commonly found in the arid regions of India, particularly in states such as Madhya Pradesh, Gujarat, Rajasthan, Western Uttar Pradesh, Punjab, Haryana, Maharashtra, Karnataka, Kerala, and in the Himalayas up to an altitude of 1500 meters.

It is typically grown by small and marginal farmers in rainfed conditions on marginal soils.

Morphology

- The plant in question is a compact, upright, bifurcated, ligneous shrub with a maximum height of 1.50 meters.
- The roots of the shrub are succulent, narrowing, and of a whitish-brown hue.
- Its ovate leaves and greenish flowers characterize this plant.

- When fully developed, the shrub produces

Fig.(19): The Ashwagandha plant is at four different stages of growth: the vegetative stage, the stage where the fruit is fresh, the stage where the root is dried, and the stage where the fruit is dried.

Climate

It flourishes in sub-tropical regions characterized by low precipitation, often grown as a long-lasting winter annual (lasting 240 days) on less fertile lands. Late winter rains are beneficial for the development of its roots.

Regions in the sub-tropics that receive between 500 to 750 mm of rainfall are deemed optimal for its cultivation. A somewhat dry season is required during its growth phase, with temperatures ranging from 20°C to 35°C being most favorable for its growth.

Soil

It flourishes in various soil conditions, with optimal growth observed in well-drained medium red and black soils. The most ideal pH level for its cultivation falls within the range of 7.5 to 8.0.

Black or heavy soils that exhibit exceptional drainage characteristics are also conducive for its growth.

Propagation:

The plant is cultivated from seeds that exhibit no dormancy period. It is advisable to select seeds of high quality from the preceding season for cultivation.

Sowing techniques

In the traditional method, seeds are usually planted directly in the field by the end of August. The line-to-line planting technique is favored for its ability to promote increased root development and facilitate efficient intercultural operations. To improve seed germination rates, it is recommended to soak the seeds in a 500 ppm Gibberellic acid solution overnight, followed by washing and treatment with Dithane M-45 (Indofil M-45) at a rate of 3 g/kg seed to protect seedlings from diseases transmitted through seeds.

Seeds are typically sown at a depth of 1-3 cm and then covered with a light layer of soil. Maintaining a line-to-line distance of 30 cm and a plant-to-plant distance of 10 cm is essential. Providing a light shower after sowing seeds can help ensure optimal germination. In certain regions, transplanting is also a common practice, where healthy seedlings aged 25-35 days are suitable for transplanting into the main field at recommended spacing.

Fig.(20): The seed of the Ashwagandha plant.

Preparation of the land

In the cultivation of Ashwagandha, the roots are considered the most economically significant component. Therefore, it is essential to properly prepare the land to ensure unhindered root development, resulting in increased length and girth, and ultimately enhancing the overall quality. The land was initially ploughed using an MB plough and harrowed twice to achieve a fine tilth following the onset of pre-monsoon rains.

During land preparation, well-decomposed farm yard manure (FYM) should be incorporated into the soil. Approximately 10-20 tonnes of FYM per hectare should be carefully mixed into the soil during the final ploughing. Subsequently, the field should be leveled through planking for optimal cultivation conditions.

Varieties

Jawahar Asgandh 20, originating from the JNKVV Regional Station in Mandsaur in 1989, is a well-known variety. Jawahar 134, released in 1998 by the AICRP on MAPs at the Mandsaur center, is another notable variety. Furthermore, the Poshita variety has been introduced by CIMAP, with Pratap and Chetak being other recognized varieties.

Organic matter and nutrients used to enhance soil fertility

Ensuring optimal crop nutrition is crucial in order to prevent any excess or deficiency of essential plant nutrients, which can have a negative impact on both the yield and quality of the produce. It is recommended to conduct soil testing prior to the application of nutrients.

Irrigation

Irrigation is necessary, it is essential to have a dependable source of clean water of high quality. Excessive water, whether from rainfall or irrigation, can be detrimental to this crop and is generally unnecessary if the monsoon season provides adequate water throughout the growing period.

Nevertheless, occasional irrigations may be required for optimal growth. When irrigation is needed, it is recommended to water the crop once every 15 days, depending on the soil composition. To preserve soil moisture, encourage proper water penetration during heavy rains, and to manage weed growth, it is advised to use organic mulches like wheat straw or the previous crop's ashwagandha straw between the rows.

Thinning and weeding

It is crucial to manage weeds before they begin to compete with the crop for essential nutrients and sunlight. A single round of hand weeding early in the growth cycle is usually adequate for allowing the ashwagandha plants to outcompete the weeds. As the ashwagandha plants mature, their smothering effect helps suppress weed growth. When performing manual weeding, caution must be exercised to avoid damaging the plant roots with the hoe.

Since the use of chemical herbicides is limited in medicinal crop cultivation, it is advisable to employ alternative weed control methods such as organic mulching to inhibit weed growth and preserve soil moisture. Two weedings are typically necessary, with the first taking place.

Harvesting

T

heir highest levels in the 180-day crop, indicating that this period is the optimal time for harvesting ashwagandha. It is advisable to harvest ashwagandha during dry weather and avoid doing so during rainy periods or early mornings when dew may be present. When harvesting, the entire plant should be uprooted carefully to avoid damaging the roots. Adequate soil moisture is necessary during harvesting to facilitate the uprooting process. Care should be taken to avoid harvesting weed plants or other foreign materials along with the crop plants.

Post-harvest handling

In order to obtain the best levels of active ingredients and superior quality, it is essential to harvest at the right time. Ashwagandha plants usually start flowering and producing fruits in December. Harvesting typically occurs between January and March, about 150 to 180 days after sowing. The maturity of the crop can be identified by the drying of leaves and the berries turning yellow-red. Research indicates that the 180-day crop is the best time for harvesting as root size, root and shoot biomass, and alkaloid content are at their peak during this period.

It is recommended to harvest ashwagandha in dry weather and avoid doing so in rainy conditions or early mornings when dew is present. When harvesting, it is important to carefully uproot the entire plant to prevent damage to the roots. Sufficient soil moisture is crucial during harvesting to aid in the uprooting process. Precautions should be taken to avoid harvesting weed plants or other foreign materials along with the crop plants.

Yield

On average, a commercial cultivation of one hectare of Ashwagandha crop can yield around 6-8 quintals of dried roots and 50-75 kg of seeds. The estimated cost of cultivating one hectare of Ashwagandha crop is 10,000, but can yield returns of 30,000 per hectare. These returns are subject to fluctuations based on the prevailing market conditions, including supply and demand dynamics.

Marketing and export

The markets of Neemuch and Mandsaur in the state of Madhya Pradesh are renowned globally for their high-quality Ashwagandha.

Climate

The plant's distribution reveals its flexibility in adapting to various climatic conditions, thriving in both tropical and subtropical regions. Nevertheless, it flourishes more abundantly in tropical climates compared to subtropical areas, where growth is hindered by cold winter temperatures.

Soil

The resilient crop thrives in a diverse range of soils, with the exception of alkaline or waterlogged ones. It thrives best in deep sandy loam to loam soils of moderate fertility, which promote root development and ease of harvesting on a large scale. The ideal pH level should not exceed 8.5.

Propagation

Plants have the ability to reproduce either through seeds or through vegetative techniques such as cuttings. When focusing on drug production, it is recommended to cultivate plants from seeds due to the quicker flowering time compared to plants propagated from cuttings. In contrast, when the goal is seed production, it is preferable to propagate plants through cuttings.

i) Propagation by seeds

It is recommended to use freshly harvested seeds for planting as their viability diminishes with prolonged storage. These seeds can be planted directly in the field or nurtured in a nursery before being transplanted as seedlings.

Direct Sowing

This technique is particularly suitable for extensive areas where labor costs are high, as it helps decrease production expenses. The land is plowed twice and brought to a fine tilth, removing any weeds, stubble, and pebbles present. The field is then divided into plots of suitable sizes, with the soil being thoroughly mixed with the recommended amount of manure and fertilizers. Seeds are sown at a rate of 2.5kg/ha during the onset of the monsoon season, typically in June or July, in rows spaced 30-45 cm apart and lightly covered.

Due to the small size of the seeds, they are often mixed with sand at a ratio of 10:1 to facilitate even distribution. Germination occurs approximately 7-8 days post-sowing.

Vegetative propagation refers to the process of asexual reproduction in plants

In order to propagate plants through this method, it has been noted that soft wood cuttings obtained from lateral shoots tend to be more successful compared to hard or semi-hard wood cuttings. Cuttings that are approximately 10-15 cm long with a minimum of 5-6 nodes are deemed to be the most effective and generally result in a rooting success rate of approximately 90%.

Immersing the cuttings in a 25 or 50 ppm NAA solution overnight has been shown to enhance rooting success even further, up to 96%. This method can be effectively utilized for increasing clones with elevated alkaloid levels and in situations where seed production is the primary objective.

Types and varieties

The roseus variant is preferred for cultivation due to its elevated alkaloid levels. Recently, the CIMAP in Lucknow has introduced two new white flowered varieties called "Nirmal" and "Dhawal," which, while equal in terms of active components, are said to produce a larger biomass.

Collecting and refining agricultural products

i. Methodical pruning of the plant's leaves, stem, and seeds is essential, starting with the initial leaf stripping at the 6-month mark after sowing, followed by a second round at 9 months. The final leaf stripping takes place after the full harvest, and the plant is then left to dry in a shaded location.

ii. Harvesting of the roots is done 12 months after sowing, with the plants cut about 7.5 cm above the ground and dried alongside the stem, leaves, and seeds. The field is irrigated properly, and once the soil is ready for excavation, the roots are carefully collected. After a thorough washing, the roots are dried in a shaded area. Seeds should be gathered from matured pods 2 to 3 months before the overall plant harvest. The upper portion of the plant, approximately 7.5 cm to 25 cm above ground level, is considered the stem for marketing purposes.

Yield

Under ideal irrigation conditions, it is possible to achieve a yield of approximately 4 metric tons per hectare of leaves, 1.5 metric tons per hectare of stem, and 1.5 metric tons per hectare of roots on an air-dried basis. In contrast, when relying on rainfall alone, the yield diminishes to around 2 metric tons per hectare of leaves, and 0.75 metric tons per hectare each of stem and roots on an air-dried basis. The leaves have a total alkaloid concentration ranging from 0.15% to 1.34%, with an average Vinblastine content of 0.002% and Vincristine content of 0.005%.

Candrabhaga, Chotachand, Serpentine root & Chandrika

For over four centuries, sarpagandha roots have been utilized for medicinal purposes, such as treating snake bites, insect stings, and nervous disorders. The plant is recognized to contain around 30 alkaloids, including important ones like reserpine, serpentine, ajmaline, ajmalicine, and rauvolfinine. Reserpine, in particular, plays a crucial role in managing hypertension in mainstream medicine. The dried roots typically have alkaloid levels ranging from 1.7 to 3%. These drugs and alkaloids derived from the plant are commonly used in mainstream medicine to treat hypertension and as sedatives or tranquilizers. The fresh roots emit a distinctive acrid aroma and have a notably bitter taste, reflecting their high alkaloid content. Harvesting of the roots mainly takes place in forested areas.

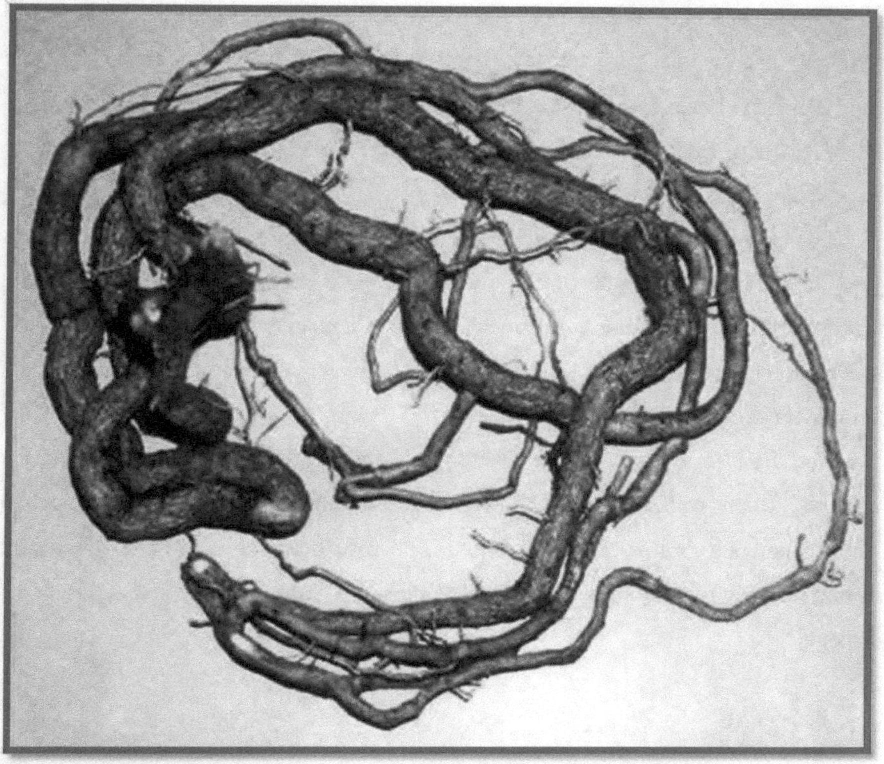

Fig.(21): Sarpagandha Roots

Source and dispersion

Sarpagandha originates from the moist, deciduous forests of Southeast Asia, including nations like Myanmar, Bangladesh, Sri Lanka, Malaysia, and the Andaman Islands. In India, it can be found in central territories like Uttar Pradesh, Bihar, Northern Bengal, specific parts of Western Ghats, and Assam.

Description of the plant

The plant is a tall perennial subshrub that maintains its evergreen qualities, typically reaching a height of 75 cm to 1 m. The leaves of this plant are characterized by their simple, elliptical shape, bright green color, and acute tips. The inflorescence is comprised of clusters of flowers arranged in corymbs, which can vary in color from white to pink.

The fruit produced by the plant is a drupe. The root system is largely tuberous, often with branching, and can reach depths of 40 to 60 cm with a diameter ranging from 0.5 to 2.5 cm. The root bark, comprising a significant portion of the overall root structure, contains a notably high concentration of alkaloids.

Fig.(22): Serpent vine flowers

Types or categories are called varieties

The College of Agriculture at Jawaharlal Nehru Krishi Vishwa Vidhyalaya in Indore launched the variety 'RS-1' with the aim of promoting its use in commercial farming.

The RS-1 strain has the ability to maintain a seed germination rate of 50-60% even when stored for a period of seven months. Additionally, the air dried root yield can reach up to 25 q/ha and contains a total alkaloid content ranging from 1.641 to 2.94%.

Soil

The plant demonstrates optimal growth in slightly acidic to neutral soils, flourishing especially well in well-drained fertile soils of medium to deep depth. An ideal pH range for this particular crop falls between 4.6 and 6.2. Clay-loam to silt-loam soils with high organic content are particularly suitable for its commercial cultivation.

Climate

Sarpagandha has the ability to thrive in diverse climates, with ideal temperatures ranging from 10 to 30 degrees Celsius. This plant flourishes in frost-free tropical to sub-tropical regions, particularly when provided with irrigation. It is most successful in areas with high levels of rainfall and well-drained soil.

Propagation

Sarpagandha has the capability to be propagated by both seeds and various vegetative methods.

Utilizing seed propagation for plant reproduction

Seed propagation is widely accepted as the most efficient technique for initiating a commercial plantation. The germination rate of Sarpagandha seeds may demonstrate notable variability, falling anywhere between 5 and 30 percent, even when exclusively dense seeds are chosen for planting. It is quite simple to differentiate between light and heavy seeds utilizing a basic water flotation method.

In the months of May and June, soaking heavy seeds in water for 24 hours resulted in a germination rate of 20-40 percent, with a recorded germination rate of 62.77 percent for freshly collected heavy seeds. Generally, 6 kg of seeds is adequate for establishing a one-hectare plantation.

Fig.(23): Serpentwood Seed

Transplanting

Seedlings that are between 40-50 days old and have 4-6 leaves are deemed appropriate for the process of transplantation. To enhance soil fertility, a generous amount of high-quality, thoroughly decomposed farmyard manure at a rate of 25-30 tons per hectare is mixed into the soil as part of the land preparation process. Subsequently, the seedlings are carefully transplanted into furrows that are dug to a depth of approximately 15 cm and spaced 45 cm apart. It is essential to ensure a consistent distance of 30 cm between each plant.

Prior to planting, the seedlings are carefully removed from the soil and the taproot is trimmed. They are then immersed in a 0.1% fungicide solution to protect against soil-borne fungi that cause damping off disease. Sarpagandha, being a slow-growing crop, requires a lengthy growth period, typically lasting 18 months or more, especially during the initial stages of growth. For a single crop, roughly 80,000 to 100,000 seedlings are necessary per hectare.

Manures and fertilizers

Seedlings with 4-6 leaves, aged 40-50 days, are deemed suitable for transplantation. During the preparation of the land, well-decomposed FYM is applied at a rate of 25-30 t/ha. The seedlings are transplanted into furrows that are dug to a depth of approximately 15 cm at intervals of 45 cm, maintaining a spacing of 30 cm between each plant. Prior to planting, the seedlings are extracted with care, their taproots trimmed, and then immersed in a 0.1% solution of fungicide to safeguard them against damping off disease caused by soil-borne fungi. Sarpagandha is a slow-growing crop, especially in its initial stages, with a lengthy growth period that spans from at least 18 months onward. As a standalone crop, approximately 80,000-100,000 seedlings are needed per hectare.

Irrigation

Sarpagandha can be cultivated as a rain-fed crop in regions with a consistent rainfall of 150 cm or more evenly spread during the growing season, such as in Assam and Kerala, under subtropical conditions. In areas where there is a significant temperature increase and low rainfall during the rainy season, regular irrigation is necessary. It is recommended to provide 15 to 16 irrigations, with watering every 20 days in the summer and every 30 days in the winter.

Intercropping

Intercropping can be successfully implemented in sarpagandha plantations, especially in areas with adequate irrigation systems. For example, soybeans can be grown alongside onions or garlic in sarpagandha plantations.

Harvesting

The roots are typically harvested 2-3 years after planting, which is around 18 months of age. It is best to harvest the roots in winter (in December) when the plants have shed their leaves, as they tend to have a higher total alkaloid content compared to roots harvested in August.

Yield

A production of 2200 kilograms per hectare of air dried roots has been achieved from a plantation that is two years old, while 3300 kilograms per hectare has been achieved from a three-year-old plantation, both grown under irrigated conditions on sandy, clay loam soil.

Dioscorea/medicinal yam, *Dioscorea floribunda*

Global diosgenin production is estimated at 1000 tonnes, with Mexico leading as the largest producer at approximately 750 tonnes annually. In India, cultivation is primarily focused in the north-eastern states and Goa, particularly through contract farming.

Plants are persistent, climbing vascular plants with tubers or rhizomes. There have been documented reports of around 600 species found worldwide, predominantly in tropical and subtropical regions and to a lesser extent in temperate zones. Among these species, only 15 are identified to possess steroidal sapogenins, particularly diosgenin.

D. composita and D. floribunda, native to Central America, along with D. deltoidea, present in North-western Himalayas, are the main species from which diosgenin is commercially harvested. These three species are classified as dioecious.

Fig.(24): Dioscorea tubers

Different forms and types of organisms

D. deltoidea **(2n=20)**

It is a native species that can be observed in its natural state in the North-western Himalayas. This plant possesses delicate and fragile vines. Propagation of this species is typically done using tuber pieces, however, the process of tuber regeneration occurs at a slow rate, requiring approximately 7-10 years for full development, even within its native temperate region. Therefore, due to the lengthy cultivation period and slow growth rate, the commercial viability of cultivating this species is not appealing to farmers.

D. floribunda **(2n=36)**

Originating from Mexico in Central America, this plant is grown in various regions including Karnataka, Goa, Assam, Meghalaya, and the Andaman Islands. The propagation of this species can be easily achieved through tuber pieces and seeds. Demonstrating dioecious characteristics, the plant is a robust climber that twines to the left.

Its branches are strong and produce compact yellow tubers at a shallow depth. The tubers contain diosgenin, with content ranging from 2-7% depending on their age. The seeds of this plant typically sprout within 3-4 weeks, leading to vigorous seedlings that thrive when transferred to the field.

Varieties

Three varieties of D. floribunda have been newly introduced for agricultural use. The strains FB (C) -1 and Arka Upkar were bred at the Indian Institute of Horticultural Research in Bangalore, whereas Pusa-1 was developed at the Indian Agricultural Research Institute in New Delhi. These cultivars are specifically adapted for tropical and subtropical environments. The distinguishing characteristics of these cultivars are outlined below:

FB (C)-1

This hybrid strain originates from cloned material imported from Central America and exhibits vigorous growth while displaying resistance to pests and diseases. The annual tuber production of this strain can vary between 20,000 to 25,000 kilograms per year when yielding 1 kilogram per plant, or up to 60,000 kilograms for a two-year-old crop producing 2.5 kilograms per plant. The dried root of this strain contains diosgenin in concentrations ranging from 2.5 to 3%.

Arka Upkar

Selected from the FB(C)-1 variety, this high yielding clone displays deeply hued, lush green foliage and demonstrates robust growth. In a crop aged two years, the tuber yield surpasses 60000 kg, with diosegenin content falling within the range of 3.5 to 4.0%.

Pusa-1

A sample chosen from germplasm displaying a tuber yield of 1.5 kg per vine within 18 months.

D. composita

This particular species is indigenous to Mexico and has been cultivated in various tropical nations.

Soil

It is capable of thriving in various soil conditions. Light or sandy soils necessitate ample irrigation and fertilization, while dense clay soils inhibit the growth and harvest of tubers due to the tendency for waterlogging. Optimal yields are achieved in loamy and deep soils abundant in organic material.

Both D. floribunda and D. composita demonstrate superior growth in red soils. It can withstand significant fluctuations in soil pH, although extreme acidity and alkalinity should be avoided.

Climate

Various types of Dioscorea plants thrive in various climatic conditions. While D. floribunda and D. composita prefer tropical climates, D. deltoidea is known to be well-suited for temperate regions. Specifically, D. deltoidea flourishes in temperate areas.

Preparation of the land prior to any construction or development

The field should undergo multiple rounds of ploughing and harrowing, be properly leveled, and have drainage channels constructed. Given the substantial need for organic matter for optimal tuber formation in yams, it is recommended to incorporate a significant amount (20-25 t/ha) of Farm Yard Manure (FYM) during land preparation.

Propagation

Yams have the ability to be propagated through various methods such as seeds, rhizome pieces, or stem-cuttings. In India, tuber-cuttings are commonly utilized for the establishment of commercial plantations. The use of seeds for propagation can lead to variability in progeny, resulting in a longer period needed for tuber production compared to plants derived from tubers.

The choice of propagating material is influenced by both the costs of planting and the unique climatic conditions of the region.

a) Propagation from tuber pieces

This particular crop thrives when grown from tuber pieces, which are typically divided into pieces weighing around 50-60 grams for optimal planting. The growth rate of plants is negatively affected and the yield is reduced if smaller pieces are utilized for planting.

There are three distinct types of pieces that can be used for planting this crop: crowns (stem end), medians (middle portion), and tips (distal ends). Crowns have the ability to sprout new shoots within 30 days of planting, whereas the other pieces may take up to nearly 100 days to germinate. The crown portion of the tubers contains a lower concentration of diosgenin compared to the median and tip portions, making them more suitable for propagation purposes. However, in cases of limited planting material availability, the median and tip portions may also be used for planting.

Irrigation

Regular irrigation is essential for the crop, particularly during the summer season.

It is recommended to irrigate the crop every 4 to 5 days in the summer and every 7 to 10 days in the winter to ensure optimal growth and yield.

Interculture

During the early stages of yam cultivation, there is a greater need for manual weeding compared to the later stages. Although chemical herbicides have been experimented with, they are not typically endorsed for commercial use. Research has indicated that it is viable to cultivate medicinal yam alongside short-duration crops such as cowpea, cluster beans, and kidney beans, particularly in areas with access to irrigation.

This method not only helps control weed growth but also enhances overall crop yield. Furthermore, intercropping can decrease the need for nitrogen fertilizers as these crops are leguminous in nature.

Harvest, output, and processing methods

The cultivation expenses are notably lower in the second year. To achieve the highest yield and diosgenin content, it is recommended to harvest D. deltoidea only after three years of growth. Harvesting of tubers typically occurs between February and March, with manual labor utilizing pickaxes for the task. Optimal yields of diosgenin are obtained when harvesting is done while the plants are dormant.

Processing

The tubers are dehydrated under the sun until they reach a moisture content of 6-7 percent, a process that typically spans a week to complete. Following this drying phase, the desiccated tubers undergo solvent extraction to separate out diosgenin

Isabgol or Blonde psyllium, *Plantago ovata*

- This medication serves as an anti-diarrheal agent.
- Its primary use is in the treatment of constipation and various intestinal disorders.
- Additionally, the seeds possess both cooling and soothing properties.
- When ingested, the seeds and husk can help alleviate inflammation of the mucous membranes within the gastrointestinal tract, including ulcers.
- Aside from its medicinal purposes, it is also employed in the fields of dyeing and calico printing.
- Furthermore, it can be utilized as a stabilizer in ice cream production.
- The seeds exhibit a range of beneficial properties.

Fig.(25): Isabgol Husk and Seed

Dispersion of goods or services

Isabgol has been introduced and cultivated in India, particularly in Gujarat and certain areas of Rajasthan. Currently, isabgol has become a major cash crop in North Gujarat, known as a 'dollar earner'.

The study of plant morphology involves examining the structure and form of plants.

This diminutive annual herb typically stands at a height of 10-15 cm and has a truncated stem. Its white flowers appear in clusters at the top of the plant, while its fruit takes the form of a capsule. Self-pollination is a common occurrence in this plant species.

The capsule, which is oval in shape and measures 8mm in length, houses smooth, dull oval seeds. These seeds are enveloped in a thin, translucent casing referred to as the husk, giving them a dark red hue and a firm texture.

Fig.(26) : Vegetative Stage

Different kinds or types of something

Gujarat Isabgol-1 and Gujarat Isabgol-2 are the two cultivars of this plant that have been introduced by Gujarat Agricultural University. An additional variety referred to as 'Niharika', which is a mutation, has been developed by CIMAP in Lucknow, known for its high yield potential.

Land preparation

The field must be devoid of weeds and clods in order to ensure optimal conditions for cultivation. The frequency of ploughing, harrowing, and hoeing should be determined based on factors such as soil quality, previous crops grown, and the extent of weed infestation.

The recommended application of farmyard manure (FYM) at a rate of 10-15t/ha should be done during the final ploughing of the field. Proper division of the field into suitable plots of varying sizes is necessary, taking into account factors such as soil texture, field slope, and irrigation requirements. For light soils with uniform terrain, a plot size of 8.0m x 3.0m would be most convenient.

Seed sowing

To achieve optimal germination rates, it is advised to utilize seeds collected from the final harvest of the previous season. The effectiveness of older seeds decreases over time when stored using conventional methods. To enhance germination success, small and lightweight seeds should be treated with a mercurial seed dressing at a rate of 3 g/kg and sown at a rate of 4-8 kg per hectare.

Before sowing, it is recommended to mix the seeds with a generous amount of fine sand or sieved farmyard manure. Broadcasting the seeds evenly is preferred over row sowing at irregular intervals, as it does not improve seed yield. Following broadcasting, lightly covering the seeds with soil by gently sweeping with a broom is advised.

It is essential to sweep the broom in a consistent direction to avoid burying the seeds too deeply, ensuring uniformity in germination. Prompt irrigation after sowing is critical, as germination typically begins around four days post-sowing. If germination is delayed, additional watering may be necessary to stimulate it.

Manures and fertilizers

During the process of preparing the land, the farmyard manure is typically applied at a rate of 10-15 tonnes per hectare. Isabgol, specifically, does not necessitate the application of large amounts of fertilizers. For optimal seed yield, a fertilizer combination of 50 kilograms of nitrogen, 25 kilograms of phosphorous pentoxide, and 30 kilograms of potassium oxide per hectare is recommended.

The complete doses of phosphorous and potassium, in addition to half of the nitrogen, are administered as a basal dose. The remaining portion of nitrogen is then divided and applied as a top dressing approximately one month after sowing.

Irrigation

Upon immediate sowing, it is imperative to provide a light irrigation. The initial irrigation should consist of a gentle flow or sprinkling of water, as a strong current may result in the displacement of seeds, leading to uneven germination and distribution.

Typically, germination of the seeds occurs within a span of 6-7 days. In the event of poor germination, a second irrigation should be administered.

Subsequent irrigations are to be applied as needed. The final irrigation should be orchestrated when a maximal number of spikes begin to emerge. A total of 6-7 irrigations are recommended for optimal crop productivity in moderately sandy soil.

The collection, transformation, and production of agricultural crops

Approximately two months after planting, the flowering process begins, and the crop generally matures for harvesting in February or March, which is around 110-130 days after planting. When the crop reaches full maturity, it acquires a yellowish tint, and the spikes themselves become brown. The seeds can be easily separated from the spikes with minimal pressure.

During harvesting, it is crucial that the environment is dry and free of moisture on the plants to prevent significant seed dispersal. As a result, harvesting should be conducted after 10 am. Following a two-day period, the crop is then threshed with the aid of a tractor in the early morning hours. To facilitate the threshing and separation process, water is sprayed over the heap.

Opium poppy/Posta/White Poppy/Afim, *Papaver somniferum*

This plant, known as an annual herb that can reach heights of 60-120 cm, is highly valued for its medicinal properties. It is a rich source of over 40 alkaloids, some of which have psychoactive effects and are utilized in psychiatry and medical research. The opium poppy plant produces a substance called 'Opium', a potent narcotic with addictive properties derived from its latex, which contains alkaloids such as morphine, codeine, and papaverine.

While the seeds of this plant do not contain alkaloids, they are known for their high linoleic acid content, which is beneficial in lowering blood cholesterol levels. Alkaloids like morphine and codeine are frequently used as sedatives for pain relief, inducing sleep, and suppressing cough symptoms. However, opium is recognized as a potent but risky substance due to its addictive nature.

These seeds are a valuable source of fatty oil and protein and are commonly used as cooking oil in many European countries. In Indian cuisine, these seeds are essential ingredients in various dishes including confectionery, pastries, and bread. Furthermore, young plants are consumed as leafy vegetables in some regions.

Uses:

- Well-established as a sedative with calming effects, it is commonly utilized for treating diarrhea and body aches.
- Opium ointment serves as a local anesthetic for conditions such as piles.
- Within Ayurveda, it is recognized for its sleep-inducing and anti-diarrheal properties.
- Its syrup is prescribed for cough relief.
- Semi-dried opium oil is employed in the production of paints and artists' ink.

Source and dissemination

The herb initially appeared in the western Mediterranean region before spreading to the Balkan peninsula, Asia Minor, and India. Over time, the cultivation of this herb has been notable in countries such as Italy, Greece, and Asia Minor. The herb was introduced to India in the 15th century. Currently, its cultivation is predominantly concentrated in the states of Madhya Pradesh, Rajasthan, and Uttar Pradesh.

Significance and molecular structure

It grows as an annual herb reaching heights of 60-120 cm. This plant holds significant medicinal value, harboring over 40 alkaloids that include psychoactive substances which have proven beneficial in the field of psychiatry for treating mental and nervous disorders, as well as aiding in medical research.

Although lacking in alkaloids, the seeds of this plant reportedly contain a substantial amount of linoleic acid which contributes to reducing blood cholesterol levels in humans. Morphine and codeine, the alkaloids present in opium, are widely utilized as sedatives for pain relief, inducing sleep, and alleviating cough symptoms. Despite its worth, opium is recognized as a valuable yet perilous substance.

In India, the primary cultivation of this plant is designated for its latex, producing opium while yielding seeds as a secondary product. These seeds provide a rich source of fatty oil and protein, commonly employed as a primary cooking oil source in several European countries.

Uses:

• Recognized for its sedative properties, opium is commonly used to alleviate symptoms of diarrhea and body pains.

• Opium ointment can serve as a local anesthetic for conditions such as hemorrhoids.

• In traditional Ayurvedic medicine, opium is utilized as a sedative and anti-diarrheal medication.

• Opium syrup is administered to treat coughing.

• A semi-dried form of opium oil is utilized in the production of paints and artist's ink.

The origins and geographical spread

The herb has its origins in the western Mediterranean region and then spread to the Balkan peninsula, Asia Minor, and India. Throughout antiquity, cultivation of this herb was popular in Italy, Greece, and Asia Minor, with its introduction to India occurring in the 15th century. Presently, its cultivation is primarily limited to the states of Madhya Pradesh, Rajasthan, and Uttar Pradesh.

Area and production

India annually produces 1,465 metric tons of opium, accounting for more than 90% of global production. The opium cultivated for commercial purposes in India is primarily concentrated in the states of Uttar Pradesh, Madhya Pradesh, and Rajasthan, spanning across 18,000 hectares.

Description of the plant

The poppy plant is a diminutive, upright herb that can reach up to 120 cm in height, boasting prominent serrated leaves and vibrant flowers in various hues such as white, pink, purple, red, and variegated. Its primary stem and offshoots culminate in sizable, oblong to globose capsules containing petite flat seeds that come in shades of whitish-yellow, grey-brown, reddish-brown, or black, and are abundant in oil content.

Fig.(26): Field of the Opium Poppy

Climate

This particular crop is commonly cultivated in regions with temperate climates, although it can also thrive during the winter months in sub-tropical areas. In order to achieve optimal growth, this crop necessitates a prolonged period of cold weather, approximately around 20°C, along with adequate sunlight during the initial growth stages to encourage robust vegetative growth; excessive rain following planting may result in diminished seed germination.

During the reproductive phase, it is imperative to have warm, arid weather with temperatures ranging between 30-35°C. Unfavorable conditions such as cloudy skies, frost, hailstorms, and strong winds, particularly during the flowering period, can cause substantial harm to the crop. Dry, warm weather during the months of February and March has been shown to improve latex flow and increase overall yields.

Soil

The cultivation of opium requires deep clay loam soils that are highly fertile and have good drainage, ideally within a pH range of 6.0 to 7.5. These types of soils, which are rich in organic matter, are able to retain moisture sufficiently for the opium plants without the need for irrigation during the lancing process. Additionally, with proper fertilization and manuring, even lighter soils such as loam to sandy-loam lateritic soils can produce high yields under optimal management practices.

Sowing

Poppy seeds should be planted in soil that has been properly prepared. The field should undergo 5-6 cross ploughings before planting takes place. It is advisable to divide the land into smaller plots to ensure efficient irrigation. Prior to sowing, it is recommended to treat the seeds with thiram in order to safeguard against soil-borne pathogens.

The ideal time for sowing is from late October to mid-November. For broadcast planting, the seed rate should be 7-8 kg/ha, while for line sowing it should be 4-5 kg/ha. Following sowing, a light irrigation should cover the seeds with a thin layer of soil. A spacing of 30x30 cm is typically maintained.

Fertilizer application

The crop necessitates essential nutrients to support the processes of flowering and capsule development. Typically, Farm Yard manure is applied at a rate of 20-30 tons per hectare by broadcasting during the preparation of the field for transplanting. Additionally, it is recommended to apply 60-80 kilograms of nitrogen and 40-50 kilograms of P2O5 per hectare. Potash is not utilized in the application. Half of the nitrogen and all of the phosphorus are applied during transplanting, with the remaining nitrogen dose administered during the rosette stage.

Irrigation

Following the initial sowing, the first irrigation should be provided promptly if the soil lacks sufficient moisture. Due to varying weather and soil conditions, a recommended irrigation schedule of 7-10 days for subsequent watering is advised. A total of 10 to 15 irrigation sessions are necessary for the successful cultivation of this crop.

Flowering and fruit-set

Approximately 90 to 100 days post-sowing, the plants that have reached waist height initiate the flowering process. Specifically, flowering typically commences in the first week of March for crops that were planted in the latter part of November. Typically, three days following the onset of flowering, the petals shed, and roughly 10 to 14 days later, the capsules are deemed suitable for lancing.

Lancing and latex collection

Lancing involves making a precise incision on the capsule, triggering the flowering of opium within 95-115 days after planting. The petals begin to fall off 3-4 days after the flowers bloom, and the capsules reach maturity 15-20 days later. At this point, the lancing process produces the most exudation. Industrial maturity is reached when the capsules display a distinctive change in color, shifting from greenish to a light green ring. A specialized knife with 3-4 equidistant points is used for lancing, ensuring the incision does not exceed 1-2 mm in depth. It is crucial to time lancing early in the morning, before 8:00 a.m., every other day to optimize results. Each incision should be one-third or less of the total length of the capsule to prevent complications.

Harvesting of seeds

After the operation of piercing the capsules and extracting opium latex, they are left to dry on the plant. The drying period typically lasts for approximately 20-25 days following the completion of lancing. In India, the capsules are manually harvested and the seeds are subsequently removed by hand by breaking the capsules open. A fully dried capsule typically weighs around 7 grams and contains between 11 to 12 thousand seeds, each weighing approximately 3.5 to 4 grams.

27

Yield of raw opium and seed

In India, an average yield of 50-60 kg/ha of raw opium and 400-500 kg/ha of seeds is typically attained.

Honey plant, *Ammi majus*

Ammi majus L., a plant belonging to the Apiaceae family, holds significance in the realm of herbal medicine. The scientific nomenclature, Ammi majus L., traces its roots back to Greek origins, with "Ammi" being derived from Dioscorides and "majus" signifying a greater size or magnitude. Believed to have originated in Egypt, Ammi majus L. thrives in sandy habitats.

The introduction of this plant in India took place at the Forest Research Institute in Dehradun in 1955, facilitated by UNESCO. Initial cultivation attempts were made in various regions of the country, particularly in Tamil Nadu, Gujarat, and Uttar Pradesh on an experimental basis. Presently, successful cultivation of this plant is being carried out in several states including Uttar Pradesh, Gujarat, Himachal Pradesh, Tamil Nadu, and Karnataka.

- It is a herbaceous annual plant capable of growing up to a height of 0.75 to 1.15 m.
- The stem, which is cylindrical, upright, and solid, features a taproot system.
- The leaves are of significant size, measuring 5-8 cm long, arranged alternately, light green in color, compound, pinnately divided, and have a lanceolate shape.
- The inflorescence takes the form of an umbel, appearing in both axillary and terminal positions, and is made up of white flowers.
- The calyx tube is fused to the bracts and features 5 teeth.

The five petals are distinct and divided. There are 5 stamens, arranged alternately with the petals. The inferior ovary is two-celled with an epigynous disk, two-lobed, and contains a single ovule in each cell.

The style is divided into two, with a minute and capitate stigma. Flowering occurs 3-4 months after sowing the seeds. The fruit is a ribbed, ellipsoid pod that is greenish-brown, turning reddish-brown upon maturity and ready for harvest within 30-35 days after flowering. The seeds are flat with thin wings, emitting a strong characteristic odor that intensifies with crushing and are slightly bitter and pungent in taste.

Fig. (28): *Ammi majus* inflorescence

Climate

Luxuriant growth of the plants necessitates moderate climatic conditions. A mild cool climate is essential during the vegetative phase, while warm and dry weather is necessary during maturity. The optimal temperature range for plant growth is between 27-32 0C. Extreme temperatures, either high or low, can negatively impact plant growth and crop yields. Regions with excessive rainfall are not conducive to the cultivation of these plants.

Soil

The crop is capable of being grown in a variety of soil types, but it thrives best in well-drained sandy loam or clay loam soils with high levels of organic matter and fertility. The land should be properly leveled with a gentle slope to facilitate drainage of excess water during periods of heavy rainfall. Unsuitable soils should be avoided for cultivation of this crop.

Nursery raising

"In well-prepared land, seed-beds of appropriate dimensions, 15-18 cm in height and with a suitable length and breadth, are established. Following seed planting, regular watering of the beds is necessary. The seeds typically begin to sprout within 10-15 days. Once germination occurs, the mulch cover can be removed. To maintain the quality of the crop, it is important to keep the seed beds free of weeds through frequent weeding."

Transplanting

The area designated for transplantation is meticulously prepared through appropriate plowing, leveling, and creation of plots of suitable dimensions. Once the seedlings are deemed to be in a sufficient state for transplanting, they are delicately uprooted to prevent any harm to their roots. The seedlings are then transplanted at specific intervals of either 45x60 cm or 45x30 cm. This process is typically carried out during the months of October and November. Following transplantation, a gentle irrigation may be administered promptly.

Irrigation

The timing of irrigation is influenced by environmental factors such as climate, soil moisture levels, and the developmental stage of the crop. Being an herbaceous plant, moisture availability plays a critical role in promoting both vegetative growth and flowering. Following transplanting, it is recommended to irrigate the field promptly and subsequently at intervals of 7-10 days, adjusting as needed based on weather conditions.

Intercultural operations

Because of the herbaceous characteristics of the crops, they are highly susceptible to competition from weeds for essential resources such as space, light, moisture, and nutrients. Therefore, timely and frequent weeding is essential to maintain a weed-free field. Additional weeding can be performed as needed. Additionally, hoeing during intervals between weeding sessions is recommended to ensure proper soil aeration.

Harvesting

The crop typically lasts for a period of 5-7 months, with flowering commencing in December. The process of flowering and the maturation of the inflorescence occur over an extended period, leading to a prolonged harvesting timeframe. If the primary umbel is not harvested promptly, there is a risk of seed shedding. It is essential to harvest the umbels when they are mature to prevent seed waste. As a result, it is advisable to harvest every 3-4 days to avoid shattering. Manual handpicking is the common method of harvesting. The harvested produce is stored in a well-ventilated, shaded area before the threshing operation is conducted.

Yield

The seed production is influenced by factors such as the variety chosen, the timing of planting, the spacing between plants, the cultural practices employed, and the timing and method of harvest. Generally, a well-managed field can yield an average of 1000-1200 kg of seed per hectare.

Belladonna/Deadly Night Shade, *Atropa belladonna*

The alkaloids atropine, hyoscine, and scopolamine are present in all parts of the plant, rendering it both toxic and hallucinogenic. The highest concentration of these alkaloids can be found in the mature fruits and green leaves of the plant. In mature fruits, atropine is the predominant alkaloid, constituting 98% of the alkaloid composition. It is estimated that each berry-sized fruit contains 2 mg of atropine. Along with atropine, scopolamine is another key alkaloid found in Belladonna plants. These alkaloids are not exclusive to Atropa belladonna, but also occur in other plants within the solanaceous family, such as Datura stramonium, Hyoscyamus niger, and Scopolia carniolica, with the latter two producing scopolamine.

The leaves of the Belladonna plant are commonly utilized in the creation of tinctures and plasters, offering various medicinal properties including pain relief, sedation, stimulation, anti-diuretic effects, anti-asthmatic effects, antispasmodic effects, and anti-inflammatory effects. They are also employed in treating renal and biliary colic, stomach complaints, and excessive sweating. The roots of the plant are predominantly utilized externally to alleviate gout, rheumatism, and other related conditions.

The Atropha belladonna plant reaches a height of approximately 1.5 meters. Its root is thick, fleshy, whitish, and branched. The calyx consists of five lobes. The leaves are simple, dark green, ovate, and arranged alternately. The berries are highly toxic. The stem is purplish in color. The bell-shaped flowers are solitary. The flowering period typically occurs from June to September. The seeds contain toxic alkaloids.

Fig.(28): Flower and Fruit

Soil

It thrives in deep, fertile soils with a medium texture that are abundant in humus. It is not suitable to grow this crop in water-logged heavy clay soils. This crop has the ability to grow in uncultivated barren lands.

Climate

Belladonna is a crop that thrives in temperate climates, displaying characteristics of a perennial and producing an optimum amount of herbage and alkaloids. In regions with sub-tropical climates, it can be cultivated as a winter crop, although it behaves as an annual due to its demise in the summer months leading to lower yields. This plant is typically found growing in the shade of trees and on hillsides.

Propagation

The propagation of the crop involves extracting seeds from berries typically gathered between September and November. Approximately 4 kilograms of seeds are needed to raise seedlings on a one-hectare plot of land. Treatment of the seeds with an 80 percent solution of sulfuric acid for two minutes during sowing can enhance germination rates.

Manures and fertilizers

Growing this crop can be physically demanding, so it is advised to use a foundational amount of 25-40 kilograms of nitrogen, 40-60 kilograms of phosphorus, and 30-50 kilograms of potassium per hectare. Furthermore, an extra 60-80 kilograms of nitrogen should be applied in 3-4 separate doses as a top dressing at monthly intervals following each harvest.

Irrigation

Belladonna exhibits a significant need for water and should be regularly irrigated every 10-15 days in dry conditions. Typically, 6-7 irrigation sessions are necessary in the dry months, with caution advised to prevent water logging.

Inter cultural operations

Belladonna should be kept free from weeds by frequent weeding and hoeing.

Harvesting

The initial yield of the leaves becomes available three months following planting. It is advisable to commence harvesting once the plants begin to flower, as this is when the concentration of alkaloids is at its peak. Pruning scissors are utilized to cut the leaves, which are then promptly dried post-harvest either in the shade, under the sun, or with the assistance of artificial heat, with or without fans to ensure adequate air circulation. While drying, it is recommended to regularly flip the leaves over. The roots, meanwhile, are harvested after a three-year growth period. They are subsequently cleaned, cut into 4-inch segments, split lengthwise if necessary, and dried in the shade or under the sun.

Yield

In the initial year, an average of 600 kg of dried herbs is harvested. The production escalates to 1500 kg per hectare in the second and third years. The quantity of dried roots harvested ranges from 170 to 335 kg per hectare.

Cinchona, *Cinchona ledgeriana*

Cinchona consists of around 65 species, among which Cinchona succirubra, C. officinalis, C. ledgeriana, C. robusta, and C. hybrida are cultivated commercially for their cinchona bark, which is the primary source of quinine and other anti-malarial drugs. In addition to quinine, more than twenty other alkaloids have been found in cinchona, with cinchonidine, quinidine, and cinchonine being the most important.

These alkaloids are mostly present as salts of quinic and cinchotannic acids, with varying concentrations depending on the species. The leaves of cinchona contain approximately 1% total alkaloids. Quinine has historically been used as an anesthetic instead of cocaine, as it has longer-lasting anesthetic effects. It has also been utilized as a sclerosing agent in the treatment of internal hemorrhoids and varicose veins.

Moreover, quinine is known for its ability to offer protection from sunburn and is acknowledged for its bitter tonic, stomachic, and appetizer properties.

Fig.(29): Profuse flowering

Fig.(30): Individual Flower of Cinchona

Distribution

It is confined to some parts of West Bengal, Tamil Nadu and Karnataka in an area of about 6000 to 8000 ha.

Soil

The plant shows a preference for a gently sloping, free-draining soil found in undisturbed forests, which is abundant in organic material and devoid of any potential for waterlogging in the deeper layers. It thrives in an acidic soil environment with a pH ranging from 4 to 6.5. The crop has a high demand for calcium in its growth and development.

Climate

It thrives most effectively in tropical environments situated at elevations of 1800 meters. Conversely, alternate species thrive in regions where the typical low temperature is13.5 degrees Celsius and the high reaches 21 degrees Celsius, with an average relative humidity of 83%. The growth is significantly hindered when temperatures fall below 7 degrees Celsius or rise above 26 degrees Celsius. Optimal growth occurs in areas where annual rainfall totals just under 200 millimeters, spread evenly over a span of at least eight months annually. Cinchona is vulnerable to frost and therefore not cultivated in extremely high mountainous regions.

Land preparation

The ideal soil type for cultivating cinchona plantations is found in virgin forest soils. Before planting, the forest area is cleared, and the land is flattened and excavated to a depth of 1½ ft to enhance soil quality and eliminate any rocks.

Approximately two weeks before planting, holes measuring 60 x 60 cm are dug and filled with topsoil and thoroughly decomposed organic material.

Propagation

Cultivation of cinchona involves propagation through both seeds and various vegetative methods such as cutting, stooling, layering, cleft grafting, veneer or side-grafting, and patch-budding. Typically, three shoots are allowed to grow for most species of cinchona, and plants are uprooted after 30 years when their vigor starts to decline.

The main harvest occurs during the first two coppicing cycles, with minimal bark yield from dead or drying trees and pruning activities. The initial yields are typically obtained in the third year post-planting, with the bark being extracted by beating it with a mallet and then peeled by hand or knife. It is crucial to dry the peeled bark promptly to avoid alkaloid loss, preferably in shaded areas or specialized sheds during rainy weather, or using artificial heat. In established plantations, well-regulated ovens are used for drying, employing hot air ovens set at 70°C for this purpose.

The long bark strips are cut into smaller pieces and fed into a rotating cylindrical oven with a slight incline for drying, resulting in a product containing around 10% moisture, which is then packed into gunny bags. This dried bark is commonly known as 'Druggists' bark in the trade, containing 1.8-2% quinine content.

Yield

During the initial two rounds of coppicing, a harvest of 4000 kg of dried stem bark per hectare is achievable. When the tree is ultimately uprooted, the bark yield may increase to approximately 6000 kg per hectare.

Fig.(30): Removal of Bark

Fig.(31): Bark after drying

Alkaloids can be obtained from the powdered bark through a process of extraction. Quinine is derived from the total alkaloid content of the bark in the form of quinine sulfate. Other alkaloids such as cinchonidine and dihydroquinine are also found in commercial products. Purification of quinine can be accomplished through multiple rounds of recrystallization to achieve a consistent specific rotation. The resulting crystalline trihydrate is a white, odorless, intensely bitter microcrystalline powder that is efflorescent.

Pyrethrum, *Chrysanthemum cinerariaefolium*

Pyrethrum is recognized as one of the most benign insecticides available, with minimal harmful effects on mammals and the ability to be metabolized if accidentally ingested. Its mode of action disrupts insects, compelling them to vacate their hiding spots, while also delivering a rapid knockdown effect. This versatile compound finds application in various industries, being utilized for the preservation of food grains, insect-resistant packaging, mosquito repellent aerosols, and coils.

Widely regarded as an effective household insecticide, approximately 70% of global pyrethrum production is directed towards the manufacturing of fly sprays and insecticidal aerosols, with another 20% dedicated to mosquito coils, and the remainder allocated to other formulations such as powders, ointments, and creams.

Fig.(32): Flowering Stage

Distribution

Kenya, Tanzania, and Rwanda collectively account for more than 80% of global pyrethrum production. Currently, commercial cultivation of pyrethrum is thriving in various regions of India, including Kashmir, Kodaikanal, and Lucknow.

Soil

Pyrethrum grows most effectively in well-drained sandy soil, although it can also thrive in red laterite loams and light to medium loam soils. Yields are notably diminished in poorly-drained soils. While the plant is capable of growing on mountain slopes and wastelands, excessively rich soils and water-logged conditions are detrimental to its growth.

Climate

It flourishes in regions characterized by a temperate, dry climate with moderate winters and cool summers. An extended period of cool temperatures, dipping below 17° C for approximately six weeks, is essential for the onset of flower-bud formation. In locations experiencing lengthy winters (from November to March) and warm summers, a solitary blooming phase occurs in May and June. Suitable growing conditions include an annual rainfall of around 1000 mm and an elevation ranging from 1500 to 2400 m.

Varieties

There are three strains of pyrethrum, namely KKL-1 which was introduced by TNAU, 'Hansa' and 'Jhelum' which were introduced by CIMAP, Lucknow for farming purposes. A new variety named 'Sel-2' with a significantly higher flower yield of 86% compared to 'Hansa' has been generated at the Kodaikanal Centre of CIMAP.

Propagation

Pyrethrum is commonly propagated through the use of seeds. It is important to collect seeds from mature plants with fully developed flowers, just as the seeds are ready to be dispersed. However, it is crucial to note that the viability of these seeds diminishes over time, so prompt usage is recommended. Alternatively, pyrethrum can also be propagated vegetatively by extracting healthy plants from the soil and using their divisions for replanting. In some regions, it is advisable to initially cultivate these divisions in nurseries before transferring them to the primary cultivation area.

Nursery raising

Seed-beds of suitable dimensions, measuring 15-18 cm in height, are established in well-prepared soil. Organic matter such as FYM or poultry manure is incorporated into the top 10 cm of the soil. The seeds are sown in rows spaced 10 cm apart, and then covered with a layer of rice-straw, wheat-straw, or dried grass. Regular watering of the beds is essential. Germination of the seeds typically occurs within 5-10 days. Once the seeds have sprouted, the mulch is removed. It is important to maintain weed-free conditions in the seed beds through frequent irrigation.

After 8-12 weeks, the seedlings will be ready for transplanting. Nurseries are typically established either in the spring (April-May) or autumn (October-November), with an extended period from March to November in irrigated regions. Optimal planting times are in the spring (March-April) or autumn (October-November), although planting can be done at any time of year if irrigation is available.

Land preparation

The land is meticulously readied through the process of 2 to 3 ploughings, subsequent leveling, harrowing, and removal of weeds and stubble. If organic manure is accessible, it is recommended to be applied during the final ploughing. For flat terrain, it is optimal to create ridges spaced 45-50 cm apart, while undulating land should be terraced for cultivation.

Planting

The seedlings are typically spaced between 27 and 45 centimeters in rows with a gap of 45 to 60 centimeters between each seedling. It is recommended to plant in ridges rather than flat surfaces as it enables better inter-cultivation and irrigation practices while also preventing water-logging.

Interculture

To achieve a successful harvest, it is essential to engage in thorough weeding and hoeing techniques. Utilizing a tractor-drawn cultivator can efficiently perform these tasks, while hand-hoes are suitable for smaller areas like terraces. The initial hoeing process can be carried out in the fall after the previous crop's harvest, after which two additional weeding and hoeing sessions should be conducted in the spring season.

Irrigation

Pyrethrum cultivation does not necessitate irrigation in instances where rainfall is evenly spread. Nevertheless, in irrigated settings, it is recommended to provide weekly irrigation during the peak growing season for optimal outcomes. In such irrigated environments, it has been noted that pyrethrum plants commence flowering after just one year as opposed to the expected two years in non-irrigated conditions. Regular irrigation is essential for the crop's sustenance during drier periods, and it is advisable to irrigate post-fertilizer application as well.

Harvesting and processing

The plants will bloom within a year after being transplanted. Initially, the crop may produce a low yield, with the optimal yields typically achieved in the second and third year.

Drying

The collected flower-heads are typically sun-dried immediately after harvesting. They are carefully laid out on straw mats and turned frequently to prevent any fermentation. During the night, they are shielded from the elements. The dehydration process typically takes 5-7 days to be fully completed. For optimal drying, hot-air drying methods are recommended, ensuring that the temperature does not exceed 80 degrees Celsius.

The weight loss during this process is usually around 65-75%. When dealing with smaller crop areas and when the risk of rain is minimal, sun-drying is a viable option. However, for larger farms, the use of air-dryers becomes necessary to ensure timely drying of the harvested flowers. Specialized mechanical drying equipment has been developed for drying pyrethrum flowers.

Yield and pyrethrin content

In Kashmir, the typical yearly crop yield is approximately 250 kg/ha, as opposed to 700 kg/ha in Kenya and 500 kg/ha in Nilgiris. Reports from Kodaikanal suggest that the initial year's yield can reach around 450 kg/ha. The overall pyrethrin concentration in India is said to range from 1.0-1.5%, with Kenya averaging 1.4% and the highest recorded at 2.1%.

Citronella grass, Java Citronella, *Cymbopogon winterianus*

Citronella oil is primarily utilized in the field of perfumery, both directly and indirectly. It is a common practice to infuse soaps, detergents, household cleaners, insecticides, and similar products exclusively with this oil for its aromatic properties. Furthermore, it serves as a valuable component in perfumery formulations for soaps and detergents. The significance of Citronellal extends to its role as a fundamental building block for the synthesis of various derivatives. Hydroxycitronellal is a derivative of citronellal that plays a crucial role in the creation of compounded fragrances.

Among floralizing perfume materials, Hydroxycitronellal is widely utilized. In the production of soap scents, a coarser grade of citronella oil is typically employed, while a higher grade is selected for flavoring compositions. Citronella oil is typically categorized into two varieties in the market:

Ceylon citronella oil, derived from Cymbopogon nardus and considered to be of lower quality, and Java citronella oil, obtained from C. winterianus and recognized as a superior variant.

Types

There are two distinct variations of this oil, categorized as Ceylon type and Java type. The Ceylon type oil is extracted from the species known as C. nardus, whereas the Java type oil is derived from the species C. winterianus. The oil obtained from the Java type is considered to be of higher quality in comparison to the Ceylon type due to its higher content of Citronellal, ranging from 30-38% as opposed to the 20-25% found in the Ceylon type.

Morphology

It is a fragrant perennial plant with fibrous roots, standing tall at over 2 meters, boasting sleek leaves, and producing a sizable cluster of flowers.

Soil

The plant thrives in diverse soil conditions, with sandy loam soil enriched with organic matter being particularly conducive to its growth. Conversely, heavy clay and sandy soils do not provide optimal support for the plant's development. The plant has been observed to flourish within a pH range of 5.8-6.0. Prolonged water retention in the soil is deemed detrimental to the plant's growth.

Climate

Citronella flourishes in tropical and subtropical climates, thriving in warm temperatures with ample sunlight and an average annual rainfall of 200-250 cm spread evenly throughout the year. Regions with high humidity are ideal for its cultivation. While the optimal altitude for growth is between 180-120 meters, the plants have been known to thrive at altitudes ranging from 1000-1500 meters. Extended periods of drought can hinder the plant's growth and development.

Fig.(33): Citronella grass

Mandakini

A cloned selection produces a slightly lower herb yield of 35 tons per hectare and 118 kilograms of oils per hectare. This variety is well-suited for cultivation in the hilly and Terai regions of the Himalayas, as noted by CIMAP in Lucknow.

Manjiri

The University of Agriculture Sciences, Bangalore has introduced this particular variety. It is a distinguished mutant clone of Manjusha M3-8, boasting 50-90% more oil and a high cirtonellol content. This variety exhibits vigorous tillering and rapid growth, leading to exceptional herb yields.

Propagation

It is a type of grass that comes back every year, but it is unable to produce viable seeds. As a result, this species can only be propagated vegetatively through rooted slips. To achieve this, well-developed clumps are carefully split, with an average of about 50 slips being produced from a one-year-old clump. Each slip contains 1-3 tillers and serves as the basic unit for propagation. Once planted, these slips grow into plants or bushes. Before planting, it is important to trim off the fibrous roots and leaves from the slips.

Season of Planting

While Java citronella plantation can be started at any point in the year, the optimal time is during the onset of the monsoon season. It is essential to prepare the land meticulously through discing and tilling to achieve a fine tilth. Adequate moisture must be present in the field during the planting process.

Method of Planting

The cuttings should be obtained from robust, healthy, and youthful shrubs, and must be promptly planted following the uprooting of the bushes and separation of the cuttings. Delays in planting may lead to partial drying of the cuttings, resulting in a diminished plant population. The cuttings should be inserted into the ground vertically, to a depth of approximately 10 cm, and spaced at intervals of 45 cm x 40 cm.

Planting should facilitate efficient drainage of excess water, as plants are highly vulnerable to transient water-logging, which can impede plant growth. To prevent water-logging, it is advisable to plant citronella on elevated ridges. In the absence of rainfall within 24 hours, the field should be irrigated immediately after planting. In the commonly employed planting method, one hectare necessitates 150,000 to 225,000 cuttings, with 2-3 cuttings per pit.

Fig.(34): Slips for planting

Manuring & Fertilization

It is a crop that depletes the soil and therefore necessitates generous use of fertilizers. The crop necessitates 10 tons per hectare of farmyard manure (FYM) and NPK at a ratio of 25:40:40 kilograms per hectare as an initial application. Typically, a substantial amount of nitrogen is needed for optimal growth. As such, 75 kilograms of nitrogen per hectare can be administered in three equal portions of 25 kilograms each at 3, 6, and 9 months after planting.

Intercultural Operations

It is important to maintain a weed-free environment in citronella plantations. Once the plants have developed into bushes, weed growth is naturally restricted due to the way the bushes block out sunlight.

However, in newly established plantations and after each harvest, weeds tend to proliferate in the spaces between rows and must be removed through weeding. An economical method for this is to use a cultivator to clear the inter-row spaces.

Irrigation

Adequate moisture is essential for optimal leaf growth and yield. In regions with an annual rainfall of 200-250 cm, evenly spread throughout the year and with high humidity, additional irrigation is typically unnecessary. During dry months, however, irrigation is required twice a week for the initial month after planting, and subsequently every five days.

Harvesting

Citronella is grown specifically for its essential oil, with the leaves containing the highest concentration of oil among all plant parts. Therefore, only the leaves should be

Harvesting at the right time is crucial as doing so too early or too late can negatively impact the oil quality and lead to decreased yield. Cutting only the leaf blade during harvesting is recommended, leaving the sheath behind as it contains minimal and low-quality oil.

Citronella plantations usually maintain their productivity for a period of 5-6 years, with the peak harvest of leaves and oil taking place in the second and third years before gradually declining. It is recommended to remove the plantation after 3-4 years and replace it with small legume species in order to promote sustainable cultivation practices.

Yield

The quantity of oil produced from the plant varies depending on the location, season, soil fertility, irrigation, management practices, and distillation method employed. Typically, the oil content constitutes approximately 1% of the total weight of the leaves. Initially, the yield of fresh leaves is around 15-20 tonnes per hectare in the first year, increasing to 20-25 tonnes per hectare in the second and third years before gradually decreasing.

In terms of oil production, the first year typically yields about 140-150 kg per hectare, while the second and third years see an increase to 200-300 kg per hectare. In the fourth year, the yield drops to 200 kg per hectare, and further drops to 100 kg per hectare in the fifth year.

Khus grass/vetiver grass, *Vetiveria zizanoides* L.

Vetiver oil is derived through the process of steam distillation of its roots. This oil is revered for its exquisite oriental scent that lingers for an extended period of time. Forming the foundation of the renowned Crepe de Chine perfume, vetiver oil, when combined with patchouli, sandalwood, jasmine, and gardenia, produces a sophisticated and complex fragrance. When utilized in blended perfumes, vetiver oil serves as a superb fixative, enhancing the longevity of volatile elements. Along with its reputation for being cooling in nature, vetiver oil is employed in the production of various products such as soaps, cosmetics, perfumes, incense sticks, beverages, and pan masala.

In terms of therapeutic benefits, vetiver oil possesses calming effects on the nervous system, aiding in the alleviation of tension and stress, and even showing promise in addressing insomnia. Widely recognized in India as the "oil of tranquility," vetiver oil can be advantageous when incorporated into baths or massage routines to address symptoms associated with conditions like arthritis, rheumatism, muscular stiffness, and discomfort. Its warming and comforting attributes can provide relief from chronic pain and help ease underlying tension. Furthermore, in skincare applications, the antiseptic and mildly astringent properties of vetiver oil can be harnessed effectively for managing oily skin prone to breakouts.

Fig.(35): Khus grass

It thrives in a moderate climate, although it can also be cultivated in both humid and arid tropical environments. This plant is remarkably resilient and can endure prolonged periods of submersion, as well as survive extreme drought conditions. It flourishes in regions with annual rainfall ranging from 100 to 200 centimeters and temperatures between 30 and 40 degrees Celsius. In colder, mountainous regions with mild winters, its growth may be limited.

Soil

It can thrive in various soil types, however, a fertile and well-drained loamy soil is recommended as ideal. Loamy soils, characterized by their loose texture and pH of 6-8, provide optimal conditions for root development and cultivation.

The plant is also capable of growing in challenging soil conditions such as waterlogged areas, sandy soils, and regions with high water tables or prone to flooding. A lush and robust root system can be achieved by cultivating the plant in warm and moist conditions on nutrient-rich, periodically inundated marshy terrain. It is advisable to steer clear of dense and compact soils.

Propagation

Propagation can be achieved through the use of tillers and slips. Tillers require a longer time to grow compared to slips, making slips the preferred planting material for propagation purposes.

Preparation of the land

The field must be cleared of any perennial weeds and shrubs before being subjected to deep tillage. Following the clearance process, the land is plowed extensively. As the pre-monsoon showers arrive, the final preparations are made for optimal cultivation. In order to ensure consistent crop growth and maximum root yield, the fields are organized into beds of appropriate dimensions.

Planting time

The period between April and August, marked by pre-monsoon showers, is widely regarded as the optimal time for engaging in commercial cultivation. To achieve maximum oil yield in South Indian climates, it is recommended to plant during the months of June and July. In regions prone to flooding, crops should be established before the onset of floodwaters to ensure protection. As such, planting should occur at least two months prior to the expected flood season.

Planting

Seedlings are carefully planted in holes that are 5-8 cm in depth, which are dug using a pointed tool. Typically, 2-3 seedlings are placed in each hole to account for possible losses and to ensure a dense population of plants. The soil around the seedlings is firmly packed and leveled to provide stability and support. The seedlings are organized in rows with a specific spacing of 45 x 30 cm for optimal growth. By employing a flat-bed planting method followed by ridging after approximately 30 days post-planting, there is an observed increase in root growth and essential oil production.

In the prevailing planting method, 1,50,000 to 2,25,000 seedlings are required per hectare, with 2-3 seedlings per hole. In case planting is carried out during a dry spell, immediate irrigation of the field is necessary post planting. While the initial growth phase may be sluggish, the plants exhibit rapid development once their roots are established.

Fig.(36): Slips for planting

The application of Farm Yard Manure (FYM) at a rate of 10 tons per hectare, along with 60 kilograms of nitrogen and 22.5 kilograms each of phosphorus (P_2O_5) and potassium (K_2O) per hectare, has proven to effectively increase the yield of vetiver.

In instances of poor soil quality, 40 kilograms each of nitrogen, phosphorus, and potassium per acre are applied prior to the final plowing. In the initial year, approximately 20 kilograms of nitrogen may be added as a top dressing around 4 months after planting, depending on soil fertility and crop growth. During the second year of crop growth, an additional 40 kilograms of nitrogen are top-dressed after rainfall in June.

The initial development of vetiver during the first 60-70 days is characterized by a slow growth rate, resulting in empty spaces between rows that are susceptible to weed infestation and the rapid growth of kharif weeds. These weeds can hinder the growth of the crop plants by competing for essential resources such as light, moisture, and nutrients. It is recommended to perform 2-3 weeding sessions at monthly intervals during the initial growth phase to manage weed growth.

Earthing up

Earthing up following the removal of weeds is advantageous for the crop. This practice promotes root development at the base of the plant by creating a more favorable rhizospheric environment for the growing vegetation.

Shoot cutting

During the onset of winter in November, the pruning of aboveground plant parts promotes the growth of tillers and leads to a higher number of roots per plant. Trimming of shoots is carried out twice throughout the cultivation period. The initial pruning occurs when the shoots reach a height of 30 cm in the first year of planting to stimulate regrowth for the following year. Subsequently, a second round of cutting is performed at a height of 15-20 cm prior to harvesting, easing the process of root extraction.

Inter cropping

Vetiver slips are typically sown during the rainy season at a spacing of 45 x 30 cm. Initially, the growth of the crop is slow, leaving the inter-row space vulnerable to weed growth. Weeds, which tend to grow rapidly, compete with the crop plants for essential nutrients and moisture. However, this open space can be efficiently utilized for cultivating intercrops without adversely impacting the main crop yield.

Harvesting

Harvesting typically takes place between 15-18 months (December-February) to ensure a fully developed root system and high-quality oil. Harvesting before the 15-month mark results in immature roots that produce oil of poor quality, characterized by a green earthy odor. Roots that have been properly developed, with a somewhat thicker composition, yield oil of superior quality, evidenced by higher values of optical rotation and specific gravity, a fuller, richer, and longer-lasting odor.

Oils obtained from older roots tend to have a darker color compared to those obtained from younger roots. If roots remain in the ground for more than two years, the oil yield significantly decreases due to the root system becoming woody, resulting in the loss of essential oil content and the production of a highly viscous, dark-colored oil that is nevertheless of high quality.

The upper portion of the plant is severed at a height ranging from 15 to 20 centimeters, and the clusters are removed from the ground. When removed using a spade or a tractor-drawn single disc, approximately 50-60% of the roots are also extracted, while the remaining roots remain in the soil.

Efforts are made to retrieve the remaining roots from the soil by further excavation. If possible, a single round of irrigation may be applied before the harvest to ease the digging process. Mechanical harvesting can be performed using a disc plough equipped with a single disc attached to a tractor, which can extract roots from a depth of 30-35 centimeters. This method not only saves labor but also boosts root recovery by around 15% compared to manual extraction. The length of the roots can range from 10 to 35 centimeters, depending on local growth conditions, soil type, and climate. Thicker roots tend to yield more oil, while very pale or white roots contain minimal oil content.

Washing/Cleaning

After being harvested, the roots are found to have substantial amounts of soil adhered to them. Following harvest, the roots undergo a gentle washing process using clean, flowing water to effectively remove the soil while being mindful to avoid losing the delicate finer roots.

Drying

The cleaned roots are laid out on designated drying areas for the purpose of dehydration. These roots are carefully rotated at specified time intervals until they reach a state of dryness. Throughout this process, any extraneous particles are meticulously removed from the collection. Once thoroughly cleaned and dried, the roots are transported to a distillery or storage area where they undergo a maturation period.

The drying procedure takes place under the shade for a duration of 1-2 days, leading to an enhancement in the aromatic properties of the essential oil. Excessive exposure to direct sunlight during the drying phase can result in a decrease in the final oil output. Following the drying process, the dried roots can be stored in a shaded environment for approximately 60-70 days without any loss in oil quantity, with a noticeable improvement in quality.

Quality aspects

The process of distillation is crucial for ensuring the quality of oil. By carefully controlling the steam pressure and adjusting post-harvest treatment of roots, both the yield and quality of the oil can be enhanced. Vetiver oil contains numerous smaller components, with the minimum level of 'Khusimol' ideally reaching 14% or higher while remaining low in caryophyllene. In addition to extraction methods, the quality of the oil is also influenced by factors such as the selection of varietals, harvesting during the appropriate age range of 15-18 months, and only during dry periods.

After the oil extraction process, the leftover roots can be utilized for crafting cartons and various handicraft items like mats, aasa (for sitting), and pen-holders. By using the water obtained during distillation, a simple syrup (sharbat) can be prepared, known for its enjoyable taste and beneficial health properties upon analysis.

Bach, Acorus, Sweet calomel, *Acorus calamus* L.

Sweet flag is indigenous to many countries in the Northern hemisphere and is commonly found throughout the United States. In India and Sri Lanka, it can be found growing naturally or intentionally cultivated at elevations reaching up to 1800 meters.

Fig.(37): Sweet Flag PlantFig.: Sweet Flag Plant

Sweet Flag is a perennial herb that thrives in semi-aquatic environments and marshy terrain. It is characterized by a creeping and highly branched rhizome, which releases a fragrant aroma. The rhizome has a cylindrical shape, measuring approximately 19-25 mm in diameter and 10 cm in length. While the exterior of the rhizome is light brown, the interior is white and spongy. The plant's leaves are robust, erect, and sword-shaped, with a robust scent when crushed. Sweet Flag blossoms with small yellow flowers arranged on a spike, although flowering and seed production are rare occurrences.

Climate

The plant is known for its resilient nature and is commonly found in regions with tropical to sub-tropical climates. Adequate sunlight is essential for the plant's growth, as well as for drying the rhizomes after harvesting. Ideal conditions for cultivation include temperatures ranging from 10°C to 38°C and annual rainfall levels between 70 and 250 cm.

The soil should be tilled two or three times before the arrival of the rainy season. The preparation of the land should mimic that of rice paddies.

Compost or Farm Yard Manure is applied at a rate of 15 metric tons per hectare in conjunction with phosphorus and potassium. The basal requirement includes one third of nitrogen, 50 kilograms of phosphorus, and 25 kilograms of potassium. The second application of nitrogen should be administered one month after planting via broadcasting, followed by a third dose two months after planting.

The riparian or canal edges where the soil is waterlogged are highly conducive for its cultivation. An initial water level of 5 cm is desired in the field, with subsequent increases to 10 cm. Irrigation may be unnecessary during the rainy season, but in the event of an extended dry period, it should be conducted every 2-3 days.

Inter-culture

It is imperative to engage in timely weeding and hoeing in order to effectively manage weed proliferation and achieve high crop yields. Following each weeding session, it is important to firmly press the growing plants into the soil.

Harvesting

After a period of 6-8 months, typically in December, the lower leaves of the plant will begin to yellow and dry, signalling that they have reached maturity. It is recommended that the field be partially dried at this stage, ensuring that there is enough moisture remaining for the plant to be uprooted. For extensive cultivation, the rhizomes can be extracted using a plough.

Post-harvest Operations

After being washed with water, the uprooted rhizome is meticulously cleaned and then trimmed into lengths ranging from 5 to 7.5 cm. At this stage, any fibrous roots are carefully removed.

Drying

The harvested rhizomes are meticulously air-dried in a shaded area to ensure that the integrity of the oil content remains unaffected.

Fig. (38):Dried rhizome

Yield

The productivity of the crop is influenced by multiple factors, including the type of variety used, spacing of plants during planting, and the agricultural methods employed. On average, the yield is estimated at around 4.22 tons of dried rhizomes or 10 tons of fresh rhizomes per hectare

Lavender, *Lavendula angustifolia* Linn.

Lavender oil emits a pleasantly pure, invigorating, and slightly sweet scent. This oil is often utilized as a replacement for authentic lavender oil and holds significant importance in the soap manufacturing sector. Due to its ability to destroy microbes, lavender oil is increasingly being utilized in aerosol form for sanitizing various spaces such as residences, educational institutions, and public venues throughout Europe. Furthermore, it is employed in medicine as a flavor enhancer and on occasion as a digestive aid. Lavender blossoms are incorporated into sachets and potpourri.

Lavender is a plant that thrives in temperate climates and is best suited for regions with cold winters and mild summers. Its extensive root system enables lavender to flourish on sloping terrain, effectively preventing soil erosion. This plant shows resilience to drought and frost, making it suitable for cultivation in high-altitude arable lands.

Soil

Light, well-drained, arid, and calcareous soils containing sufficient nutrients are the most optimal for cultivating lavender. While poor carbonate and sandy soils are also acceptable, soils that retain excess water are not suitable for growing lavender. Studies indicate that lavender produces greater yields of herbs and oils when grown in neutral to alkaline soils with a pH level ranging from 7.0 to 8.4.

Lavender has the ability to propagate itself through two methods: seed propagation and vegetative propagation. While seed propagation is cost-effective and efficient, it can lead to a wide range of variations in generation forms due to cross-pollination. This complexity can present challenges in the cultivation and harvesting of lavender at optimal times.

The seeds are planted in nursery beds during the autumn months of November and December. A quantity of 0.2-2.5 grams of seeds is necessary for every square meter of area. The seeds should be buried at a depth of 1-2 centimeters. Germination of the seeds occurs in the springtime, typically in April, when temperatures are around 14-15 degrees Celsius. It is essential to regularly trim the seedlings to prevent the formation of follicles and ensure consistent growth of the young plants.

Fig.(39): Seed

Fertilizers

Typically, a foundational application consisting of 20 kilograms of nitrogen, 40 kilograms of P_2O_5, and 40 kilograms of K_2O is utilized prior to planting. Additionally, 80 kilograms of nitrogen per hectare is administered in intervals of four separate doses throughout the annual cycle.

Irrigation

During periods of drought, it is necessary to maintain regular irrigation of the crop in order to achieve a satisfactory spike yield.

Harvesting

Once the flowering process has commenced, the plants are carefully gathered for harvesting. The entire plant is collected, ensuring that the bottom one-third is left untouched to allow for natural regrowth. The freshly harvested herbage is utilized for the process of steam distillation.

Extraction of essential oil

The plant material undergoes steam distillation for the extraction of essential oils. Various varieties of the herbage contain essential oil content ranging from 0.5% to 1.1%, with an average content of 0.8%.

Yield

On average, third-year plantations produce 2000 kg of vegetation per hectare.

Rose geranium, Scented Geranium, *Pelargonium graveolens*

It is a significant aromatic plant known for producing an essential oil highly valued for its intense and distinct rose-like fragrance. The genus Pelargonium comprises approximately 600 species, some of which offer pleasing scents. However, species such as P. radula and P. fragrance are less notable and lack commercial importance.

Uses

The pure essence of geranium oil possesses a strong aromatic quality that stands out on its own and harmonizes effectively with various other scents. It is frequently employed in the fragrance of soaps and in the extraction of rhodinal, which is a key component in many premium perfumes. To satisfy the needs of the Indian perfumery sector, India is importing in excess of 20 tonnes of geranium oil from foreign sources, supplementing an annual domestic production of roughly 20 tonnes.

Types

1. The Algerian or Tunisian variety of geranium features slender, erect plants bearing dark pink flowers and contains 0.3% oil with 57% 1-citronellol. Cultivated in the Nilgiris region, this type is ill-suited for damp conditions but produces a significantly higher oil yield with a more delicate fragrance compared to the Reunion variety.

2. In contrast, the Reunion or Bourbon type of geranium displays a more bushy growth habit and light-pink flowers, with oil comprising a greater concentration of

1-citronellol (59.4%). Grown in the Nilgiris and Anamalais regions, this variety thrives in moist conditions.

Plant Description

It is a densely branched, fragrant plant. The stem is cylindrical, becoming woody at the lower portion, covered in fine hairs, initially green and later turning brown. The leaves are arranged alternately on the stem, stipulate, simple, with five primary lobes and secondary lobes, and densely covered in fine hairs. The leaves emit a strong aroma. The inflorescence is in a clustered umbel shape and covered in fine hairs. The flowers are bisexual, with a pink corolla, and positioned beneath the ovary. There are ten stamens with nearly equal-length filaments that are fused at the base; seven anthers are present. The ovary is covered in fine hairs, positioned above the flower, with five divided sections and fused carpels; the style is hairy, breaking at the end into five stigma sections

Fig.(40) Scented geranium **a)** Vegetative stage **b)** Flowering stage

Climate

The crop can be cultivated in a range of climates, including temperate, subtropical, and tropical regions, at elevations between 1000 and 2200 meters. Optimal growth occurs in subtropical environments with temperatures between 5 and 23 degrees Celsius, with temperatures below 30 degrees Celsius being harmful to the plant.

Soil

The crop in question has a shallow root system, necessitating well-drained soil with good porosity. It thrives in red lateritic soils with a pH ranging from 5.5 to 8.0, but soil rich in calcium and with good drainage is ideal for optimal growth.

Propagation

Propagation of geraniums is easily accomplished through cuttings, as this plant does not produce seeds for reproduction. The most effective method involves using terminal cuttings that are approximately 20 cm long and contain about 8 nodes, as they demonstrate an 80% rooting rate even without any treatment. Middle and basal cuttings, on the other hand, tend to show poor rooting capabilities, but this can be improved by applying growth regulators such as IBA or IAA at 200 ppm for a duration of 6 minutes. Planting the cuttings in raised beds that are 3 m long and 1 m wide, with well-mixed soil containing powdered FYM, is recommended. The cuttings should be placed closely together with a spacing of 8-10 cm, and their cut ends should be dipped in a 0.1% Benlate solution before planting. Providing temporary shade and regular watering before root initiation, as well as bi-weekly spraying with a 0.2% Urea solution, will prepare the cuttings for transplanting. Rooting the cuttings in polythene bags offers additional protection for the root system before planting in the main field, ultimately leading to a higher success rate.

Planting

Approximately 30,000 cuttings are necessary for the cultivation of one hectare of land. Prior to planting, the land must be adequately prepared through ploughing and brought to a suitable condition for growth. Ridges and furrows are created, with fertilizer and irrigation being applied the day before planting takes place. The cuttings are carefully selected from the nursery and placed at intervals of 60 cm x 60 cm.

Varieties

Hemanti, Bipuli, and Kunti have been introduced by the Central Institute of Medicinal and Aromatic Plants (CIMAP) in Lucknow specifically for cultivation in the plains of Northern India. Additionally, KKL-1, Kelkar, Ooty, and IIHR Sel-8 are among the alternative varieties that can be found for this particular crop.

Manures and Fertilizers

When applied, the crop respons positively to both major and micronutrients. For soil incorporation, well-decomposed farmyard manure at a rate of 15-20 tons per hectare, along with 35 kg of nitrogen, 35 kg of phosphorus pentoxide, and 35 kg of potassium oxide per hectare, are utilized in the form of urea, super phosphate, and muriate of potash.

Harvesting

It is typically harvested 4 months after being transplanted, at which point the leaves start to take on a light green hue and undergo a transition in scent from lemon-like to that of rose. However, this process necessitates close attention and hands-on experience. Harvesting should be done with a sharp sickle and promptly sent for distillation. The utilization of a sharp sickle is crucial to minimize jolts, pulls, and harm to the crop during harvesting. Following each harvest, the crop should be hoed, fertilized, and irrigated according to a set timetable. Subsequently, the plant will produce new shoots, grow at a faster rate, and be ready for the next round of harvesting within 4 months. In this manner, a total of 3 harvests can be achieved over a span of 3-6 years. The utilization of polyhouse conditions in cultivation has been shown to shorten the harvesting time by 21 days. The essential oil is mainly found in the green parts of the plant, especially the leaves. Oil levels are generally higher during the summer months, typically from April to June. The top portion of the plant with 6-12 leaves contains a greater amount of oil compared to the middle and base sections.

Distillation of Oil

The newly harvested plant terminals are utilized in the process of oil distillation. The plant material is arranged near the distillation apparatus for a period of approximately 12 to 24 hours. This period of rest allows for a slight fermentation process to occur, causing the glycosides to split and resulting in an increased oil yield. The oil itself is then extracted through a straightforward distillation technique.

The plant material is densely packed into the still above a perforated grid, tightly compressed, and the still-head securely closed. Steam is produced in a distinct boiler and directed towards the still. As the steam rises, carrying the oil vapors along, it is subsequently condensed by passing through a condenser chilled with running cold water. The condensed oil is then separated from the water using a method based on their differing density, and further refined by filtering it through activated carbon.

Patchouli, Pehpli (hindi), ***Pogostemon patchouli***

The plant is a perennial herbaceous shrub that produces aromatic leaves that contain highly fragrant essential oil.

The extracted essential oil finds applications in perfumery, toiletry, and tobacco industries. It is especially sought after for its effectiveness as a fixative in high-end perfumes, providing strength, character, alluring notes, and long-lasting qualities. Highly prized for its aromatic properties, the essential oil is utilized in perfumes, soaps, cosmetics, and flavor industries.

Distribution

It is grown in the coastal areas of Tamil Nadu, Karnataka, Assam, and West Bengal. Cultivated over an area of approximately 600 hectares, this crop yields 20 tons of oil annually.

Soil

The plant is known for its resilience and ability to thrive in various soil types. Nevertheless, it thrives best in a deep loamy soil that is enriched with humus and essential nutrients, has a loose friable texture, and lacks an impervious layer at the bottom to ensure optimal growth and production. It is crucial that the pH level of the soil falls within the range of 5.5 to 6.2.

Climate

The ideal climate for this crop is warm and humid. It thrives when cultivated in regions with consistent rainfall between 150-300 cm per year. Additionally, the crop can be successfully grown with irrigation in areas with lower levels of precipitation. Optimal growing conditions include a temperature range of 24-28 oC and an average atmospheric humidity of 75%. Furthermore, this crop can flourish at altitudes ranging from 800 to 1000 meters above sea level.

Varieties

The important varieties are Johore, Singapore, Java, Malaysia and Indonesia.

Propagation

The propagation of the plant is achieved through stem cuttings consisting of 4-5 nodes with a length of approximately 15-20 cm. To initiate growth, the initial 2-3 pairs of leaves on the cuttings are delicately removed and then placed in nursery beds with a spacing of 2-3 cm between each cutting. The beds are meticulously watered by hand and shaded for optimal conditions. Typically, the cuttings develop roots within 4-5 weeks and are prepared for transplantation in approximately 8-10 weeks. Alternatively, earthen pots and polythene bags can be utilized for nurturing the nursery stock.

Irrigation

The field undergoes irrigation immediately following transplantation to ensure proper growth and survival of the plants. Adequate shade and moisture are crucial during the initial stages of plant development. Subsequently, the irrigation frequency is adjusted based on the soil's water retention capacity and prevailing weather conditions. In the initial days post-transplantation, daily irrigation is necessary for 3 to 4 days, followed by alternate day irrigation for 10 to 15 days. Depending on soil type and climate, irrigation may be required once or twice weekly for a three-week period. It is important to note that the crop is vulnerable to waterlogging.

Shading

The plant thrives in shaded areas and has been effectively cultivated as an undercrop in arecanut and coconut orchards in Kerala under irrigated conditions. Additionally, it may be cultivated by intercropping with appropriate shade trees. To ensure adequate shading, Gliricidia or Erythrina trees can be planted at a spacing of 5 x 5 meters in patchouli fields.

Harvesting

The initial harvest can be conducted approximately 5 months post-planting. Subsequent harvests may occur every 3-4 months, contingent upon the level of care provided during the growth period spanning 3-4 years. Optimum timing for harvesting involves selecting cool hours to prevent any essential oil loss, targeting leaves that have transitioned in color but have not yet reached yellow or brown. The harvesting process entails harvesting young shoots measuring between 25-50 cm in length, each possessing a minimum of 3 pairs of matured leaves.

Curing and yield

After being harvested, the stem and leaves are carefully arranged in a single layer on a sturdy, dry surface in sheds to undergo the curing process, ensuring adequate air circulation. The drying phase typically lasts around three days, with the quality of the resulting oil heavily dependent on this step. To ensure uniform drying and prevent fermentation, it is essential to regularly flip the material during this stage. Once fully dried, the material is compacted into bales and stored in a cool, dry location until it is ready for distillation. A successful harvest typically produces approximately 2 tonnes of dry leaves annually, resulting in roughly 50-60 kg of oil per hectare.

Pudina, *Mentha sp.*

Fig.(41): Mentha plant

The oil extracted from spearmint is commonly incorporated into perfumes and various cosmetic products such as scents, soaps, aftershaves, and colognes. Carvone is the primary component of spearmint oil, comprising 57.71% of its composition, while other minor constituents include phellandrene, limonene, L-pinene, and cineole. Spearmint oil is predominantly utilized as a flavoring agent in toothpaste, as well as in food products like pickles, spices, chewing gum, confectionery, soaps, and sauces.

Japanese mint has the capability to thrive in tropical and subtropical regions with the help of irrigation, but it is vulnerable to root-rot during damp winters. Optimal vegetative growth occurs at temperatures ranging from 20-25°C; however, research suggests that the production of essential oil and menthol increases notably at higher temperatures, around 30°C, particularly in Indian conditions. Unlike Japanese mint, peppermint and spearmint struggle to achieve profitable growth in tropical and subtropical regions, particularly in areas with excessively high summer temperatures reaching up to 41°C. The most favorable conditions for the highest yield are found in humid and temperate climates such as those in Kashmir, Uttar Pradesh, and Himachal Pradesh. A vital factor for successful growth and oil development is access to open, sunny locations without excessive rainfall during the growing season. Bergamot mint, on the other hand, has the potential to thrive in both temperate and subtropical regions, with higher yields generally observed in temperate climates.

Soil

Paragraph 1: Soil that is moderately fertile and rich in humus is optimal for growing mint. It is important that the soil has good water retention capabilities, while also preventing waterlogging. The ideal pH range for mint cultivation is between 6 and 7.5.

Propagation

It spreads by way of creeping stolons or suckers, with peppermint and bergamot mint also being propagated through runners. The optimal period for harvesting stolons is typically from December to January.

Improved Cultivars

(A) Japanese mint

- Himalaya (MAS0-1): This particular variety, developed by CIMAP Lucknow, boasts an oil content ranging from 0.8 to 1.0% and a menthol content of 81%.
- Kalka (Hybrid 77): CIMAP Lucknow has bred this tall and robust variety known as Kalka.
- Shivalik: Introduced from China and subsequently released by CIMAP, Lucknow.
- EC-41911: This selection is a result of crossbreeding between M. arvensis and M. piperita.

(B) Peppermint

- Kukrail: CIMAP has developed and released this high-yielding peppermint variety.

(C) Spearmint

- MSS-1: A cultivar of spearmint introduced from the USA, this selection was released by CIMAP.
- MESS-5: This variety, derived from MSS-1, was also released by CIMAP.

• Punjab spearmint-1: A clonal selection, with varieties like Arka and Neera recently introduced by CIMAP.

(D) Bergamot mint

• Kiran

Other varieties include Gomti, Kosi, Saksham, and Kushal.

In order to cultivate mints effectively, it is essential to prepare the soil by thoroughly ploughing and harrowing it until it is finely textured. Prior to planting the crop, all weeds and stubble must be carefully removed. Mints are typically cultivated on either flat land or ridges, necessitating the creation of appropriately sized beds or ridges as per the recommended spacing.

Fig.(42): Propagating material

Manuring can be carried out during the land preparation phase by incorporating Farm Yard Manure at a rate of 25 to 30 tonnes per hectare. Prior to planting mint, green manuring can also be conducted. Sun-hemp is considered to be an excellent choice for a green manure crop. Mint cultivation demands a significant quantity of nitrogenous fertilizers in order to enhance the yield of herbage. Typically, an application of nitrogenous fertilizers at a rate of 80-120 kg/ha, phosphorus at 50 kg/ha, and potassium at 40 kg/ha is necessary to achieve a healthy crop yield.

The water demand of the mint crop is significant and varies according to the soil and climate conditions. The crop typically needs to be irrigated 7-10 times before the onset of the first monsoon. Following the monsoons, the crop necessitates three irrigation sessions in September, October, and November. In some cases, an additional irrigation may be necessary during the winter season if the plant is in a dormant state and there is a lack of rainfall to promote growth of the underground stems. In temperate regions, mint cultivation requires only 3-4 irrigation sessions from July to October.

Regular maintenance, such as weeding and hoeing, is necessary during the initial phases of crop growth. Weeding should be conducted frequently within the first 75 days after planting. The presence of weeds can lead to a significant decline of up to 60% in herb and oil yields.

Crop rotations are utilized to effectively manage weed growth, sustain soil fertility, and optimize yields from agricultural land. In Uttar Pradesh, key crop rotations include sequences such as mint-maize-potato, mint-early paddy, and potato, and mint-late paddy and sweet pea.

The Japanese mint is typically ready for harvesting after a period of 100-120 days post-planting, when the lower leaves begin to display a yellow hue. Timely harvesting is crucial to prevent leaf loss and oil depletion. It is advisable to harvest in clear, sunny weather conditions. The harvesting process involves cutting the green herb using a sickle approximately 2-3 cm above the ground. A subsequent harvest can be obtained around 80 days after the initial harvest, followed by a third harvest after a similar period of time. In contrast, peppermint, spearmint, and bergamot mints cultivated in temperate regions typically yield their first crop by the end of June and a second in September or October.

The newly harvested herb has an oil content of 0.4%. An abundant harvest of Japanese mint has the potential to yield as much as 48 tons per hectare of fresh herb. Nevertheless, the average yield of mints from three harvests is between 20 to 25 tons per hectare. The average oil yield ranges from 50 to 70 kilograms per hectare. While both bergamot mint and Japanese mint have similar average oil yields of 70 to 100 kilograms per hectare, peppermint oil has a lower average yield of 50 kilograms per hectare.

Distillation of Oil

Mint oil is derived through the distillation process of the fresh herb. This distillation process can be carried out using both traditional and modern distillation equipment. In the traditional method, water and steam-distillation principles are utilized, whereas in the modern approach, steam from a separate boiler is used. Before distillation, the stems are carefully removed from the dried material due to their high proportion (30 to 50%) in the material and minimal oil content.

Musk/Muskdana, *Abelmoschus moschatus*

The upright botanical herb, originating from India, is commonly referred to as Latakasturi in Ayurveda, Kattu Kasturi in Siddha, and Ambrette in English. Noted for its non-toxic nature, this plant features edible leaves, flowers, and seeds, with the latter two being consumable in their raw form. Each component of this herb holds medicinal value and finds application in various practices.

The plant is a tall, annual herb adorned with coarse hairs, graciously reaching heights of up to 2 meters above nutrient-rich soil. Its leaves are intricately lobed with extended petioles, showcasing vibrant flowers nestled in the upper leaf axils. These flowers boast petals of yellow hue accentuated with profound purplish markings near the base. The plant's seeds, seed oil, and concentrated extract are commonly exchanged within commercial markets.

The cultivation of ambrette crop in the subtropical regions of India is distributed in small areas. Annual production of ambrette seed oil is below 1000 kg. Key producers of this crop include India, Colombia, Ecuador, New Guinea, Northern Australia, Columbia, and Brazil.

The aromatic oil derived from this plant is commonly utilized in the production of perfumes, cosmetics, and various scents. It contributes a musky fragrance to items such as sachets, pan masala, and incense sticks. A significant portion of the seeds harvested from this plant is distributed to European nations.

It is necessary to use a soil turning plough for one pass and two to three passes with a soil pulveriser to ensure proper cultivation. Achieving proper leveling is recommended and can be achieved through the use of planking. Incorporating farmyard manure into the soil is an important step during land preparation.

Propagation and Seed Rate

The propagation of this plant occurs through seeds. The recommended seed rate is 1.5kg per hectare. The seeds should be sown in a field that has been thoroughly prepared. They should be sown in rows at a depth of 1 cm using dibbling techniques. Germination typically begins within 4-15 days after sowing, with completion occurring within 15-30 days, provided that conditions are favorable. The ideal temperature for germination is approximately 30°C. The germination rate for high-quality commercial seed is around 85%.

The crop has a high demand for fertilizers. Farmyard manure is incorporated into the soil during field preparation at a rate of 5-6 tons per hectare. A combination of 40, 30, and 30 kilograms of nitrogen, phosphorus, and potassium, respectively, is applied before planting, with additional 40 kilograms of each nutrient applied 40 and 90 days after germination.

Harvesting

The flowering of its plants begins in October and can last until the end of March. Harvesting of the pods should be done on a weekly basis when three-fourths of them turn blackish brown. This process can be quite challenging due to the presence of hairs on the plants and pods that can cause irritation. In India, the harvesting season typically ends by March as the later rounds of harvesting result in minimal yield that is not economically viable. Furthermore, the seeds must be dried in the shade to achieve the desired moisture level before they are stored.

Tulsi (Hindi), Ajaka, Manjari (Sanskrit), Thulasi (Tamil), Sacred Basil, Holy Basil, *Ocimum sanctum L.*

The biennial shrub known as Ocimum sanctum Linn., commonly referred to as the 'Sacred basil' or 'Holy basil', belongs to the Lamiaceae family and is frequently cultivated in gardens. This species holds a sacred significance for Hindus in India and is traditionally cultivated in courtyards and temples. The leaves of this plant, when subjected to steam-distillation, produce a volatile oil with a bright yellow color and a pleasant aroma that is characteristic of the plant, accompanied by a noticeable hint of cloves.

Its leaves, seed, and root also have medicinal properties. The leaves are rich in ascorbic acid and carotene. The leaf juice is known to have diaphoretic, antiperiodic, stimulating, expectorant, and antipyretic effects, making it beneficial for conditions such as catarrh and bronchitis, as well as for skin ailments like ringworm. Additionally, drops of the leaf juice can help alleviate earaches.

The plant is a tall, perennial herb with an upright growth habit. Its branches are numerous and covered in soft hairs. Standing at a height of 30-75cm, this plant displays elliptic-oblong leaves that are either pointed or rounded, with smooth or serrated edges. The leaves are densely pubescent on both surfaces, with tiny glandular dots scattered throughout. The plant produces purplish or crimson flowers arranged in close clusters known as racemes.

It flourishes in a wide range of soil types, from fertile loam to impoverished laterite, from saline and alkaline to moderately acidic soils, all of which are conducive to its cultivation. Adequately draining soils contribute to enhanced vegetative growth, while waterlogged conditions can lead to root rot and hindered growth.

Raised seed-beds of adequate dimensions should be meticulously prepared by incorporating farmyard manure (FYM). Approximately 200-300 grams of seeds are required to produce seedlings for planting one hectare of land. The seeds should be planted at a depth of 2 centimeters in the nursery beds. Following the seed sowing in the nursery, a mixture of FYM and soil should be evenly spread over the seeds and watered with a sprinkler system. Germination typically occurs within 8-12 days, and the seedlings are usually ready for transplantation after approximately 6 weeks, when they have reached the 4-5 leaf stage. Application of a 2% solution of urea on the nursery plants 15-20 days before transplanting can contribute to the production of robust and healthy plants for transplantation.

The gathered crops are typically distilled in their natural state soon after harvesting. Nevertheless, the quality and quantity of oil do not decrease for up to 6-8 hours following harvest, but any additional delay could result in significant reductions in both yield and quality. It has been determined that steam distillation is more effective than water distillation. The entire herb contains essential oil ranging from 0.1 to 0.23%. The quantity of oil extracted varies depending on the variety, time of year, and geographical location. The estimated yield of oil is between 10-23 kg per hectare.

The extraction, utilization, and economics of pharmaceuticals and aromatic plants

Essential oils are utilized in a diverse array of consumer products, including cosmetics, pharmaceuticals, perfumes, soaps, detergents, toiletries, confectionery, soft drinks, distilled beverages, and insecticides. The global production and consumption of essential oils and perfumes are experiencing rapid growth. Improving production technology is crucial in enhancing the overall yield and quality of essential oils. Traditional processing methods for essential oils remain significant and continue to be utilized in various regions worldwide. Water distillation, water and steam distillation, steam distillation, cohobation, maceration, and enfleurage are among the most traditional and commonly employed techniques.

Maceration proves to be advantageous in cases where distillation fails to produce an adequate amount of oil. Distillation techniques are successful in isolating oils from ground almonds, rose petals, and rose blooms, whereas solvent extraction is favored for valuable, fragile, and heat-sensitive materials like jasmine, tuberose, and hyacinth. The method of water distillation is highly regarded for the production of citronella oil from plant matter.

Sources of natural essential oil

In general, essential oils are derived from a variety of plant components including flowers (such as rose, jasmine, tuberose, carnation, clove, rosemary, and lavender), leaves (like mint, Ocimum spp., lemongrass, and palmarosa), as well as leaves and stems (such as geranium and patchouli), bark (like cinnamon, cassia, and canella), wood (including cedar, sandal, and pine), roots (such as angelica, sassafras, vetiver, and valerian), seeds (like fennel, coriander, caraway, dill, and nutmeg), fruits (including bergamot, orange, lemon, and juniper), rhizomes (such as ginger, calamus, and curcuma), and gums or oleoresin exudations (such as balsam of Peru, Myroxylon balsamum, storax, myrrh, and benzoin).

There are four main techniques employed for the extraction of essential oils from plants: distillation, enfleurage, maceration, and solvent extraction. The selection of a specific method is determined by various factors, such as the characteristics of the plant material, the amount of oil content present, and the overall stability of different plant components.

1. The process of distillation, which is the primary means of extracting essential oils, necessitates careful consideration at the outset. It comprises the technique by which the elements of one or more liquids are separated based on disparities in their vapor pressure.

2. Three prevalent distillation methods are as follows:

a. Hydro or water distillation

b. Wet steam distillation, which incorporates water and steam

c. Direct steam distillation

3. Water distillation entails the submersion of the plant material or unrefined oleoresin in water within an enclosed apparatus, subsequently subjected to heating to the boiling point through direct application of heat or perforated steam coils.

4. The advantages of water distillation include:

It allows for the processing of finely ground material or botanical components. In steam distillation, powdered materials clump together, preventing steam from properly penetrating them. The water distillation apparatus is cost-effective, simple to assemble, and well-suited for use in various field settings.

Complete removal of material is unattainable as certain esters undergo partial hydrolysis and aldehydes have a tendency to polymerize. The process necessitates a large number of stills, additional space, and more fuel due to the inability to completely vaporize high boiling and water soluble constituents. This inefficiency makes it uneconomical.

When utilizing water steam distillation, the plant material is placed on a grid above the water level within the still. Steam is then generated and directed into the water, creating 'wet' steam that passes through the aromatic plant material at low pressure.

The advantages of water steam distillation include a higher yield of oil, less decomposition of the oil, as well as reduced fuel and time requirements. Furthermore, the quality of oil produced through steam and water distillation is more consistent and reproducible. This method is also faster than traditional water distillation, resulting in improved energy efficiency. It is worth noting that many essential oils are currently produced using steam and water distillation methods, such as lemongrass oil in Bhutan, where a rural steam and water distillation system is employed.

Disadvantages:

• The method is not suitable for materials that tend to clump together, like rose and orange blossoms. The process necessitates a considerable amount of steam to completely vaporize the high boiling range oil due to the low pressure of the rising steam.

• The plant material may become damp, leading to a delay in distillation as the steam first needs to vaporize the water to allow further condensation up the still.

• To prevent the lower plant material from becoming saturated with water and to avoid excessive boiling, a baffle is utilized to impede direct contact between the water and plant material on the grid.

C) Direct steam distillation involves the introduction of live steam into the base of the still, passing through aromatic material placed on grids at intervals. This method is known for its efficiency and speed. It is particularly beneficial for the hydrolysis of ester, if required, and results in a higher oil recovery rate.

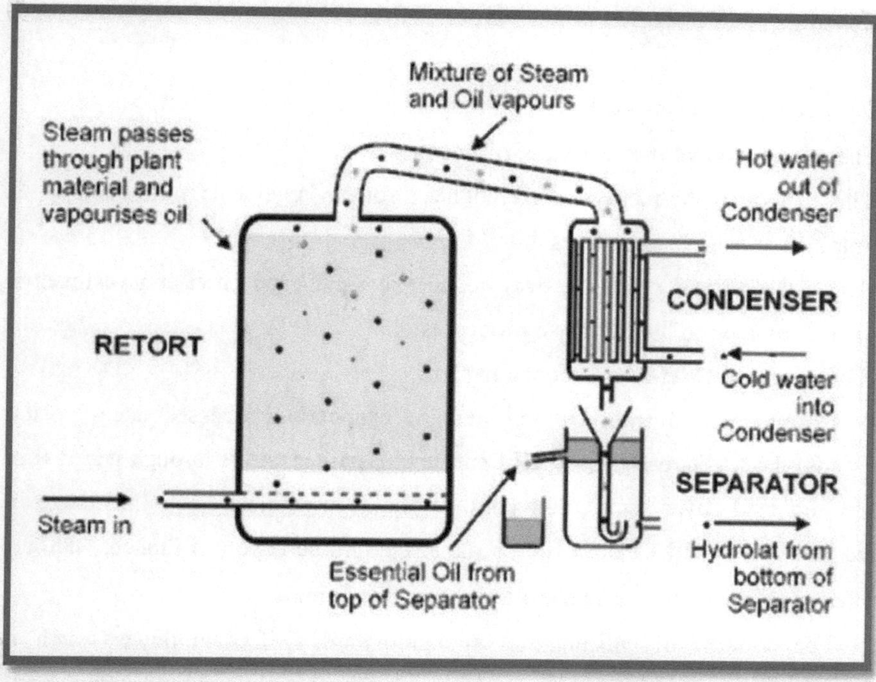

Fig.(43) : Direct Steam Distillation

Advantages of Direct Steam Distillation

• The quantity of steam can be easily regulated.

• There is no thermal breakdown of oil components.

• This method is the most widely recognized for the large-scale production of oil, surpassing alternative processes.

• The establishment of this activity requires a significantly greater investment in capital compared to the other two processes.

2. Enfleurage or Cold-Fat extraction method

Despite the advancements in modern extraction processes using volatile solvents, the traditional technique of enfleurage, passed down through generations and perfected over time, continues to hold significance. Enfleurage is predominantly practiced on a large scale in the Grasse region of France, with rare instances in India where it remains in its primitive form.

The concept of enfleurage is straightforward. Certain flowers such as tuberose and jasmine retain their perfume-producing abilities even after being harvested. Each jasmine and tuberose flower can be likened to a small factory, releasing tiny quantities of perfume continuously. The high absorption capability of fat allows it to readily absorb the emitted perfume when in contact with fragrant flowers, forming the basis of enfleurage.

The enfleurage process involves spreading freshly harvested flowers on a specially prepared fat base (known as corps) for a period of 24 hours in the case of jasmine and longer for tuberose. These flowers are then replaced with fresh ones throughout the eight to ten week harvest period, gradually saturating the fat with flower oil. At the conclusion of the harvest, the flower oil is extracted from the fat using alcohol and isolated for use.

The efficacy of enfleurage hinges largely on the quality of the fat base utilized. Precision and diligence are essential in the preparation of the corps, which must be virtually devoid of scent and possess the correct texture. If the corps is too firm, the flowers may not adequately interact with the fat, diminishing its absorption capacity and leading to a lower yield of flower oil. Conversely, if the corps is too soft, it may overwhelm the flowers, causing the exhausted ones to stick; upon removal, the flowers may retain some fat, resulting in significant shrinkage and corps loss. Therefore, the corps should exhibit a semi-hard surface to facilitate easy removal of the exhausted flowers. Enfleurage is conducted in cool cellars, and manufacturers must adjust the corps preparation to match the cellar temperatures during the flower harvest season. Years of experience have shown that a blend of one part highly refined fat and two parts lard is ideal for enfleurage, ensuring the corps has appropriate consistency and absorption capabilities. The resulting fat corps is white, velvety, consistently uniform, free of moisture, and essentially odorless. Certain manufacturers may incorporate small amounts of orange flower or rose water in the corps preparation, mostly for traditional purposes. These additions subtly impact the scent of the final product, imparting a delicate hint of orange blossom or rose.

In summary, this technique is valuable for extracting intricate floral fragrances from flowers that continue to generate essential oils even after harvesting (such as jasmine, tuberose, violets, etc.) or in cases where the oil is impacted by high temperatures. Within this approach, a thin layer of refined fat is applied to both surfaces of a glass plate stacked one on top of the other to create an airtight chamber within a structure known as a 'chassis' constructed in a cool, dark environment or basement. Fresh blooms are positioned atop the fat layer. At regular intervals, the spent flowers are substituted with a new set until, after approximately four weeks, the fat becomes thoroughly infused with the scent. The flowers can be removed either mechanically or with the aid of vacuum devices.

Maceration

Certain plant materials need to be soaked in warm water in order to release their essential oils, as their volatile components are bound by glycosides. An example of this is the wintergreen leaves which contain gaultherin and primeverosidase; when these leaves are soaked in warm water, the enzyme primeverosidase acts on the gaultherin to release methyl salicylate and primeverose.

Other examples include brown mustard (sinigrin), bitter almonds (amygdalin), and garlic (alliin). In this method, batches of chopped plant material are heated with hot oil at temperatures between 45-80o C for several hours. The filtered mixture is then mixed with fresh flowers in successive batches, up to 20 times. The resulting oil is then subjected to alcoholic extraction to obtain the essential oil.

4. Solvent extraction

When natural-flavored oil is needed, the process of direct extraction using solvents like petroleum ether or benzene is commonly used. Initially, the solvent is passed slowly through the flowers to extract the essential oils and waxes. This 'solvent' is subsequently evaporated under a vacuum, resulting in a semisolid residue of essential oils and waxes in the retort. The residue is then treated with alcohol to dissolve the essential oil, with the waxes either filtered out or precipitated through freezing. Subsequently, the alcohol can be evaporated to obtain the floral absolute.

Methods for preserving essential oils

Aromatic crops are cultivated for the purpose of extracting their essential oils, which are acquired using a variety of extraction techniques. The value of these essential oils is heavily influenced by their quality. The evaluation of quality entails an examination of the physical and chemical characteristics of the oils.

During the steam distillation process, a combination of distilled water and essential oil is gathered in the receiver of the distillation apparatus. Most essential oils possess a lower density than the distillation water, causing them to rise to the surface. The essential oil is then extracted from the receiver, filtered to remove impurities like ash, debris, and plant material, and treated with anhydrous sodium sulfate to eliminate moisture. Subsequently, the oil is stored in a cool, dark environment, filled to capacity in containers with secure closures to maintain its quality.

Adhering to proper cleaning and storage protocols is vital in preserving the quality of essential oils over time. Research has shown that exposure to water, air (oxygen), and sunlight can have detrimental effects on the quality of essential oils during storage, with water having a more pronounced impact than light.

The implementation of effective storage methods is essential for safeguarding the quality of essential oils from production until they are marketed or utilized in consumer products like flavorings and fragrances. Inadequate storage conditions can lead to alterations in the composition of essential oils over time. To ensure the quality of both crops and essential oils during storage, specific guidelines must be followed.

1. It is recommended that preserved dried crops be stored in a dry, well-ventilated structure that maintains consistent daily temperatures and proper air circulation.

To prevent infestation by pests and wildlife, it is advisable to have air conditioning and humidity control systems in place, as well as wire screens on openings to deter intruders.

2. The storage area should have a clean, crack-free floor that is easy to clean and maintain. Plant material should be stored on shelves at a distance from walls to avoid issues like pest infestation, mold growth, and oil loss. Regular inspections should be carried out to protect against these risks, and it is suggested that preserved dried crops be stored in a building with concrete floors, away from walls, and adequately separated from other crops.

3. Quality control measures must be implemented consistently during all processing stages to ensure the use of standard materials free from contaminants and foreign matter. Processed medicinal plant materials should be packaged according to the specific regulations of the producer's country and the end-user countries, using clean, non-polluting packaging materials that meet quality standards for the respective medicinal plant materials.

4. Fragile medicinal plant materials should be packaged securely in robust containers. Dried medicinal plants, herbal drugs, and essential oils should be stored in a well-ventilated facility with stable temperatures. Fresh medicinal plant materials should be stored at low temperatures, ideally between 2-8°C, whereas frozen products should be kept below -20°C.

5. Small quantities of crude drugs should be stored in airtight, moisture-proof, and light-proof containers like tin cans, metal tins with lids, or amber glass containers. It is not recommended to store crude drugs in wooden boxes or paper bags.

Marketing of Medicinal and aromatic plants

The pharmaceutical industry has evolved globally, accompanied by advancements in chemical techniques, leading to the substitution of crude drugs with pure chemical drugs. This shift has led to a decline in the use of medicinal plant-based therapy, particularly in developed countries.

The growing demand for natural medicine is closely associated with the increasing popularity of environmentally-friendly consumption practices. Medicinal plants offer various benefits, such as simplicity, efficacy, a wide range of therapeutic properties, and a focus on preventive care rather than curative treatments. According to the World Health Organization (WHO), more than 80% of the global population relies on traditional plant-based medicine for their primary healthcare needs.

In the global market, Medicinal and Aromatic Plants (MAPs) fall under the category of "Herbs and Spices" and function within the same framework of trade and distribution as herbs and spices. There are only a limited number of dealers and brokers who specialize exclusively in medicinal herbs. Over the past few years, there has been a noticeable rise in direct trading between producers and exporters in developing countries (particularly medium to large-scale operations) and processors in consumer nations. The successful trading of MAPs relies heavily on the establishment of mutual trust and confidence between suppliers and processors. Major herb processors and prominent trading companies maintain close relationships with their main suppliers through site visits and provide technical guidance on matters such as harvesting and production. The international marketing of this sector is primarily steered by importers, processors, wholesalers, and retailers.

The herbal industry in India is composed of 8610 licensed herbal units, as well as unregulated cottage-level operations, and a multitude of traditional healers and household users of herbal materials. This industry is interconnected with a intricate trade network that transports herbal materials from various sources to consumers.

India has a long history of incorporating traditional healthcare practices dating back to the 4th millennium BC. The healthcare sector in India heavily relies on medicinal plants, with over 95 percent sourced from the wild and less than 20 species being commercially grown. The global market value for Medicinal and Aromatic plants has significantly increased, reaching USD 3 billion in 2015 from USD 1.1 billion in 1999 and USD 1.8 billion in 2009. India is home to a diverse plant life, with 17,000 species of flowering plants, 8000 of which are used for medicinal purposes. India is the second largest exporter of medicinal plants in terms of volume and value.

Medicinal and Aromatic plants collected from forests or cultivated land can be distributed through various market channels in different forms, including raw materials, semi-processed materials, and consumer products. The supply chain typically involves collectors selling their products to local, road-head, regional, and country-level traders. Aromatic plant products undergo a similar distribution process, with collectors selling to processors who extract essential oils for sale to overseas traders and manufacturers. Major buyers of medicinal plants include Indian companies and pharmaceutical firms, who typically purchase from traders or intermediaries.

The demand for Medicinal and Aromatic Plants (MAPs), particularly those cultivated in high-altitude mountainous regions exceeding 3000 meters, is experiencing a notable increase on a global scale. There is a surging interest among consumers worldwide in eco-friendly products, specifically health remedies, flavors, and scents. Professionals in the field of science, including researchers and multinational pharmaceutical corporations, are actively investigating natural compounds derived from plants as potential remedies for a variety of illnesses, as the efficacy of synthetic drugs has proven to be somewhat limited. MAPs are indispensable in the pharmaceutical sphere for the creation of various formulations such as tablets, capsules, powders, and extracts. Additionally, they are utilized in a wide range of industries including Ayurveda, Traditional Chinese medicine, pharmaceutical research, aromatherapy, perfumery, health supplements, and dye production.

There are significant obstacles associated with trading Medicinal and Aromatic Plants (MAPs) that must be addressed at both the domestic and international levels.

Collectors, processors, and traders face challenges when it comes to identifying suitable markets due to various factors such as pricing, quality, and quantity considerations. Local and regional price fluctuations, coupled with a lack of timely market and pricing information, can impact the income of local collectors who often rely on intermediaries or village traders. Financial and logistical constraints can also impede the ability of local producers and collectors to cultivate closer relationships with potential clients.

The lack of knowledge and incomplete data regarding products, markets, and pricing among collectors and producers presents a significant obstacle. In addition, the difficulty in accessing appropriate markets and relevant market information further complicates the situation. There is a noticeable absence of recognition for environmentally and economically sustainable Medicinal and Aromatic Plants (MAPs), and the inability to personally meet with traders or buyers and build relationships with them only exacerbates the issue.

Complications in policy and regulatory frameworks related to trading MAPs can hinder business operations within this sector. Furthermore, the absence of standardization and consistency in quality standards for international trade adds another layer of complexity to the challenges faced by collectors, processors, and traders.

Rigorous phytosanitary regulations, including mandates for microbiological, pesticide, and heavy metal evaluations imposed by global markets. Disparities in the distribution of advantages within the sector.

Limited presence of regulated MAP markets, predominantly situated in major urban centers.

Markets for MAPs are difficult to access, with distances exceeding 50km for local markets and more than 100km for herbal mandis, resulting in heightened transportation expenses.

Scarcity of cooperatives or Agricultural Produce Market Committees (APMCs) to oversee the marketing of medicinal plants. Absence of a minimum support price for farmers involved in MAP production.

Exporting MAPs requires securing CITES authorization, a protracted and cumbersome procedure that frequently deters farmer interest in exporting their harvests.

The lack of technical competence leads to inadequate grading of cultivars, causing producers to not receive fair prices commensurate with the quality of their products. Instead, producers often receive a single payment for their entire yield.

Proposed strategies for enhancing the market infrastructure of Medicinal and Aromatic Plants (MAPs) encompass:

- The establishment of co-operatives among farmers cultivating Medicinal and Aromatic Plants (MAPs) is essential in order to ensure producers are informed about market trends and conditions.
- Local communities and farmers should be educated on the importance of "Access and Benefit sharing" to safeguard traditional knowledge about plants and prevent exploitation of this valuable asset.
- Collaboration with institutions and universities is necessary to expand extension programs focused on MAP cultivation, providing producers with scientific support for their crops.
- Detailed planning is required for the sustainable collection, cultivation, and utilization of medicinal plants to preserve genetic diversity.
- Prioritizing research and development in technology is essential for improving the cost-effective production of Medicinal and Aromatic Plants (MAPs).
- The South Asian region is rich in a variety of medicinal and aromatic plants, encompassing herbs, shrubs, trees, and vines that thrive in vulnerable ecosystems primarily inhabited by rural and indigenous communities.
- While these plants hold significant medicinal and economic value, they are currently facing threats due to inadequate conservation efforts and unregulated trade practices that exploit natural resources.
- Effective management of MAP resources and strict enforcement of trade regulations have the potential to promote biodiversity conservation and provide vital resources to rural communities for sustainable livelihood development.

Conclusion

Valuable observations have been made regarding the extensive history of utilizing medicinal plants in the southern Fars province, underscoring the profound bond between indigenous communities and their surrounding ecosystem. The thorough cataloging of 171 plant species, including their various uses and the traditional knowledge associated with them, greatly enhances the expanding collection of ethnobotanical resources. Nevertheless, there are specific aspects that warrant additional inquiry in order to comprehensively grasp and safeguard this esteemed biocultural legacy:

Species identification: Accurate identification of plant species that were not collected due to limitations is imperative for the exploration of their potential medicinal properties and for contributing to a comprehensive record of the medicinal flora in the region.

Subspecies of Teucrium polium L.: Given the documented variations in medicinal uses within Teucrium polium L., further investigation is needed to study its subspecies and their unique chemical compositions in order to understand the reasons behind these diverse applications.

Phytochemical analysis: Traditional medicinal plants that have not undergone a thorough assessment of their chemical constituents require phytochemical examinations to validate their therapeutic potential and identify bioactive compounds. This is particularly important for species such as Haplophyllum laristanicum C.C.Towns., a native plant known for its significant medicinal properties.

By addressing these knowledge gaps, researchers can gain a deeper understanding of the array of medicinal plants in the region, which can support the development of sustainable conservation strategies and potentially lead to the discovery of new sources of medicinal agents.

This research emphasizes the importance of combining traditional ecological knowledge with scientific research to ensure the lasting preservation of both biological and cultural diversity.

Throughout the annals of history, individuals have strived to discover solutions to alleviate pain and combat various illnesses. As civilizations have advanced, the healing properties of certain plants have been identified, recorded, and passed down through the ages. This enduring fascination with medicinal plants has transformed into contemporary approaches to their cultivation and application. Our study sheds light on the role of traditional herbal medicine in urban areas of Casablanca, underscoring its importance in oral health maintenance for the local populace. The information gathered has the potential to preserve ancestral wisdom regarding the utilization and preparation of diverse botanical species.

The research, conducted among traditional healers in a densely populated region of North Africa, underscores the ongoing preference for plant-based remedies in addressing oral conditions, particularly gum ailments. The trust that the community places in traditional herbal medicine can be attributed to its efficacy and affordability in comparison to modern alternatives. In Morocco, traditional healing practices continue to be widespread, representing a valuable legacy that merits protection. Phytotherapy should be recognized as a legitimate form of therapy for all individuals, necessitating further exploration into documented plants for ethnopharmacological investigations to validate traditional knowledge with scientifically supported therapeutic solutions.

While contemporary agriculture heavily leans on synthetic chemicals, apprehensions regarding their safety have prompted the investigation of low-risk, eco-friendly biocontrol substitutes. Botanical substances like essential oils and phenolic compounds are being harnessed for pesticide-free pest management due to their biodegradable properties and precise targeting. Despite the expanding market for biopesticides, their efficacy and performance assessments are limited, impeding their widespread availability for commercial use. The inclusion of aromatic herbs in agricultural systems can enhance crop variety and soil fertility, with resulting by-products functioning as organic fertilizers to suppress soil-borne pathogens. Adopting aromatic herbs in pest control programs can reduce the environmental impact of chemical pesticides.

The growing concern over the waste produced by the aromatic and medicinal plant industries has resulted in a focus on repurposing by-products for natural biopesticides and organic fertilizers. Utilizing composting and vermicomposting techniques provides effective methods for producing stable end products that are rich in beneficial compounds and microorganisms, ultimately improving soil quality and nourishing crops.

Botanicals, particularly essential oils, have been extensively studied and widely used as treatments to combat soil-borne plant pathogens. There is compelling evidence suggesting that many of these botanicals possess strong antifungal properties, which can inhibit the growth of various fungal species such as Fusarium, Aspergillus, Penicillium, Rhizoctonia solani, Verticillium dahliae, and Sclerotinia sclerotiorum. Furthermore, they have been found to exhibit significant nematicidal effects against root-knot nematodes (Meloidogyne spp.).

Despite their potential benefits, the utilization of these treatments is hindered by various limitations, including insufficient in vitro studies, inconsistent selectivity, a variety of bioactive compounds with complex mechanisms of action, rigorous approval processes, and registration requirements. Therefore, it is crucial to gain a deeper understanding of these challenges and find ways to address them in order to improve the commercialization of plant-based biocontrol agents.

References

1. Davidson-Hunt I. Ecological ethnobotany: stumbling toward new practices and paradigms. MASA J. 2000;16:1–13.
2. Liu Y, Dao Z, Liu Y, Long C. Medicinal plants used by the Tibetan in Shangri-la, Yannan, China. Ethnobiol. Ethnomed. 2009;5 doi: 10.1186/1746-4269-5-15.
3. Kebriaee-zadeh A. Overview of national drug policy of Iran. Iranian J. Pharm. Res. 2003;2:1–2.
4. Rechinger KH. Flora Iranica. Academische Druch-u, Verlagsanstalt, Graze-Austria. 1976;120:15–16.
5. Mozaffarian VA. A Dictionary of Iranian Plant Names. Tehran : Farhang Moaser; 2006. pp. 198–515.
6. Rout SD, Panda T, Mishra N. Ethno-medicinal plants used to cure different diseases by tribls of Mayurbhanj district of north Orissa. Ethnomed. 2009;3:27–32.
7. Prance GT, Chadwick DJ, Marsh J. Ethnobotany and the Search for New Drugs. Chichester : Wiley J and Sons Ltd; 1994. p. 44.
8. Zargari A. Medicinal Plants. 6th ed. Vol. 3. Tehran : Tehran University Publication;
9. Khalighi-Sigaroodi F, Jarvandi S, Taghizadeh M. Therapeutic Indications of Medicinal Plants. Tehran: Arjmand; 2010. pp. 1–6.
10. Fleming T. PDF for Herbal Medicine. New Jersey: Medical Economics Company; 1998. pp. 604–1217.
11. Khare CK. Indian Medicinal Plants: an Illustrated Dictionary. Berlin: Springer; 2007. pp. 10–49.

12. Faghir MB. Botanicals, a Phytocosmetic Desk Reference. Guilan University Press; 2005. pp. 354–355.
13. Amin GH. Popular Medicinal Plants of Iran. Tehran: Tehran University of Medical Sciences; 2005. pp. 38–162.
14. Alimirzaee P, Gohari AR, Hajiaghaee R, Mirzaee S, Jamalifar H, Monsef-Esfahani HR, Amin Gh, Saeidnia S, Shahverdi AR. L-methyl malate from Berberis integerrima fruits enhances the antibacterial activity of ampicillin against Staphylococcus aureus. Phytother. Res. 2009;23:797–800. doi: 10.1002/ptr.2641
15. Jamshidzadeh A, Niknahad H. Hepatoprotective activity of Berberis integerrima Bge. extract in rats treated with CCl4: in-vitro and in-vivo studies. Toxicol. Let. 2006;164
16. Ziai SA, Rezazadeh Sh, Dastpak A, Shabestari A, Taghizadeh M, Naghdibadi HA, Poorhoseini L. Study of the ACE inhibitory effect of medicinal plants used in Iranian folk-medicine as antihypertensive remedy. J. Med. Plants. 2006;5:53–74.
17. Zargari A. Medicinal Plants. 6th ed. Vol. 1. Tehran : Tehran University Publication; 1997. pp. 249–265.
18. Bown D. Encyclopedia of Herbs and their Uses. London: Dorling Kingersley; 1995. p. 253.
19. Ghorbani A. Studies on pharmaceutical ethnobotany in the region of Turkmen Sahra, north of Iran (Part 1): General results. J. Ethnopharmacol. 2005;102 doi: 10.1016/j.jep.2005.05.035
20. Mahasneh AM. Screening of some indigenous Qatari medicinal plants for antimicrobial activity. Phytother. Res. 2002;16:751–3. doi: 10.1002/ptr.1037.

21. El-Darier SM, El-Mogaspi FM. Ethnobotany and relative importance of some endemic plant species at El-Jabal El-Akhdar Region (Libya) World J. Agric. Sci. 2009;5:353–60.
22. Iranian Ministry of Health and Medical Education, Deputy Ministry for Food and Drug. Iranian Herbal Pharmacopoeia. Tehran: Iranian Ministry of Health and Medical Education; 2002. pp. 161–695.
23. Heidari MR, Azad EM, Mehrabani M. Evaluation of the analgesic effect of Echium amoenum Fisch and C. A. Mey. extract in mice: possible mechanism involved. J. Ethnopharmacol. 2006;103:345–9. doi: 10.1016/j.jep.2005.08.027.
24. Ranjbar A, Khorami S, Safarabadi M, Shahmoradi A, Malekirad AA, Vakilian K, Mandegary A, Abdollahi M. Antioxidant activity of Iranian Echium amoenum Fisch and C. A. Mey flower decoction in humans: a cross-sectional before/after clinical trial. Evidence-based Compl. Alter. Med. 2006;3:469–73. doi: 10.1093/ecam/nel031. [DOI] [PMC free article] [PubMed] [Google Scholar]
25. Shafaghi B, Naderi N Tahmasb L, Kamalinejad M. Anxiolytic effect of Echium amoenum L. in Mice. Iranian J. Pharm. Res. 2002;1:37–41.
26. Musto CJ. The Ancient and Medieval Pharmaceutical Treatments for Arthritis, Gout, and Sciatica [dissertation] Raleigh: North Carolina State University; 2009.
27. Mirhidar H. Plants Learnings. 5th ed. Vol. 6. Tehran : Daftare Nashre Farhange Eslami; 2003. p. 364.
28. Zargari A. Medicinal Plants. 6th ed. Vol. 2. Tehran : Tehran University Publication; 1996. p. 619.

29. Tosun A, Bahadir O, Altanlar N. Antimicrobial activity of some plants used in folk medicine in Turkey. Turk. J. Pharm. Sci. 2006;3:167–76.

30. Sonboli A, Eftekhari F, Yousefsadi M, Knani MR. Antibacterial activity and chemical composition of the essential oil of Grammosciadium platycarpum Boiss. and Husskn. From Iran. J. Biosci. 2005;60:30–34. doi: 10.1515/znc-2005-1-206.

31. Ebrahimzadeh MA, Nabavi SM, Nabavi SF. Correlation between the in-vitro iron chelating activity and poly phenol and flavonoid contents of some medicinal plants. Pak. J. Biol. Sci. 2009;12:934–8. doi: 10.3923/pjbs.2009.934.938.

32. Sayyah M, Moaied S, Kamalinejad M. Anticonvulsant activity of Heracleum persicum seed. J. Ethnopharmacol. 2005;98:209–11. doi: 10.1016/j.jep.2004.12.026.

33. Hajhashemi V, Sajjadi SE, Heshmati M. Anti-inflammatory and analgesic properties of Heracleum persicum essential oil and hydroalcoholic extract in animal models. J. Ethnopharmacol. 2009;124:475–80. doi: 10.1016/j.jep.2009.05.012.

34. Narayanan S, Ruma D, Gitika B, Sharma SK, Pauline T, Sai Ram M, Ilavazhagan G, Sawhney RC, Kumar D, Banerjee PK. Antioxidant activities of seabuckthorn (Hippophae rhamnoides) during hypoxia induced oxidative stress in glial cells. Mol. Cell. Biochem. 2005;278:9–14. doi: 10.1007/s11010-005-7636-2.

35. Suleyman H, Buyukokuroglu ME, Koruk M, Akcay F, Kiziltunc A, Gepdiremen A. The effects of Hippophae rhamnoides L. extract on ethanol-induced gastric lesion and gastric tissue glutathione level in rats: a

comparative study with melatonin and omeprazole. Indian J. Pharmacol. 2001;33:77–81.

36. Zeb A. Important therapeutic uses of sea buckthorn (Hippophae): a review. J. Biol. Sci. 2004;4:687–93.
37. Zargari A. Medicinal Plants. 6st ed. Vol. 4. Tehran: Tehran University Publication; 1997. p. 103.
38. Azaizeh H, Saad B, Khalil K, Said O. The state of the art of traditional Arab herbal medicine in the Eastern region of the Mediterranean: a review. Evidence-based Comp. Alter. Med. 2006;3 doi: 10.1093/ecam/nel034.
39. Rivera D, Obon C, Inocencio C, Heinrich M, Verde A, Fajardo J, Llorach R. The ethnobotanical study of local Mediterranean food plants as medicinal resources in Southern Spain. J. Physiol. Pharmacol. 2005;56:97–114.
40. Pieroni A, Muenz H, Akbulut M, Baser KHC, Durmuskahya C. Traditional phytotherapy and trans- cultural pharmacy among Turkish migrants living in Cologne, Germany. J. Ethnopharmacol. 2005;102 doi: 10.1016/j.jep.2005.05.018.
41. Amiri H, Khavari-Nejad RA, Masoudi Sh, Chalabian F, Rustaiyan A. Composition and antimicrobial activity of the essential oil from stems, leaves, fruits and roots of Smyrnium cordifolium Boiss. from Iran. J. Essent. Oil Res. 2006;18:574–77.
42. Amiri H, Khavari-Nejad RA, Rustaiyan A. Chemical composition of essential oil and the study of secretory anatomy from Smyrnium cordifolium Boiss. Pajouhesh Va Sazandegi. 2007;74:11–16.
43. Esmaili A, Amiri H. The study of quantitative and qualitative changes of essential oil from Smyrnium cordifolium Boiss. in Lorestan province. J. Med. Plants. 2006;5:36–41.

44. Rabbani M, Sajjadi SE, Zarei HR. Anxiolytic effects of Stachys lavandulifolia Vahl. on the elevated plus-maze model of anxiety in mice. J. Ethnopharmacol. 2003;89 doi: 10.1016/j.jep.2003.09.008.
45. Safaei A. Identification and quantitative determination of Luteolin and Apigenin in the aerial parts and an extract of Stachys lavandulifolia by HPLC. Iranian J. Pharm. Res. 2004;3:90–90.
46. Dhar U, Manjkhola S, Joshi M, Bhatt A, Bisht AK, Joshi M. Current status and future strategy for development of medicinal plants sector in Uttaranchal, India. J. Curr. Sci. 2002;83:956–64.
47. Ozturk S, Ercisli S. Antibacterial activity and chemical constitutions of Ziziphora clinopodioides. J. Food Cont. 2007;18:535–40.
48. Shuaib, M. et al. Ethnobotanical and ecological assessment of plant resources at District Dir, Tehsil Timergara, Khyber Pakhtunkhwa, Pakistan. *Acta Ecol. Sin.* **39**, 109–115 (2019).
49. Awan, A. A., Akhtar, T., Ahmed, M. J. & Murtaza, G. Quantitative ethnobotany of medicinal plants uses in the Jhelum Valley, Azad Kashmir, Pakistan. *Acta Ecol. Sin.* **41**, 88–96 (2021).
50. Idm'Hand, E., Msanda, F. & Cherifi, K. Ethnobotanical study and biodiversity of medicinal plants used in the Tarfaya Province, Morocco. *Acta Ecol. Sin.* **40**, 134–144 (2020).
51. Bahadur, S. et al. Traditional usage of medicinal plants among the local communities of Peshawar valley, Pakistan. *Acta Ecol. Sin.* **40**, 1–29 (2020).
52. Mahwasane, S. T., Middleton, L. & Boaduo, N. An ethnobotanical survey of indigenous knowledge on medicinal plants used by the traditional healers of the Lwamondo area, Limpopo province, South Africa. *South. Afr. J. Bot.* **88**, 69–75 (2013).

53. Tembo, N., Lampiao, F., Mwakikunga, A. & Chikowe, I. Ethnobotanical survey of medicinal plants used for cervical cancer management in Zomba District, Malawi. *Sci. Afr.* **13**, e00941 (2021).

54. Hosseini, S. H., Sadeghi, Z., Hosseini, S. V. & Bussmann, R. W. Ethnopharmacological study of medicinal plants in Sarvabad, Kurdistan province, Iran. *J. Ethnopharmacol.* **288**, 114985 (2022).

55. Hosseini, S. H., Bibak, H., Ghara, A. R., Sahebkar, A. & Shakeri, A. Ethnobotany of the medicinal plants used by the ethnic communities of Kerman province, Southeast Iran. *J. Ethnobiol. Ethnomed.* **17** (2021).

56. Sadeghi, Z., Kuhestani, K., Abdollahi, V. & Mahmood, A. Ethnopharmacological studies of indigenous medicinal plants of Saravan region, Baluchistan, Iran. *J. Ethnopharmacol.* **153**, 111–118 (2014).

57. Sadat-Hosseini, M., Farajpour, M., Boroomand, N. & Solaimani-Sardou, F. Ethnopharmacological studies of indigenous medicinal plants in the south of Kerman, Iran. *J. Ethnopharmacol.* **199**, 194–204 (2017).

58. Ghorbani, A. Studies on pharmaceutical ethnobotany in the region of Turkmen Sahra, north of Iran (Part 1): General results. *J. Ethnopharmacol.* **102**, 58–68 (2005).

59. Ebrahim Hosseini, S., Tavakoli, F. & Karami, M. *Medicinal plants in the treatment of diabetes Mellitus*. *J. Clin. Exc* **2** (2014).

60. Mehrnia, M., Akaberi, M., Amiri, M. S., Nadaf, M. & Emami, S. A. Ethnopharmacological studies of medicinal plants in central Zagros, Lorestan Province, Iran. *J. Ethnopharmacol.* **280**, 114080 (2021).

61. Delfan, B., Bahmani, M., Rafieian-Kopaei, M., Delfan, M. & Saki, K. A review study on ethnobotanical study of medicinal plants used in relief of

toothache in Lorestan Province, Iran. *Asian Pac. J. Trop. Dis.* **4**, S879–S884 (2014).

62. Wangchuk, P., Yeshi, K. & Jamphel, K. Pharmacological, ethnopharmacological, and botanical evaluation of subtropical medicinal plants of Lower Kheng region in Bhutan. *Integr. Med. Res.* **6**, 372–387 (2017).
63. Khan, M. N. & Badshah, L. Floristic diversity and utility of flora of district Charsadda, Khyber Pakhtunkhwa. *Acta Ecol. Sin.* **39**, 306–320 (2019).
64. Khan, W. M., Shah, S. Z., Khan, M. S. & Akhtar, N. Evaluation of indigenous knowledge of medicinal plants from Tall Dardyal Hills, Khyber Pakhtunkhwa, Pakistan. *J. Herb. Med.* **20**, 100314 (2020).
65. Aziz, M. A., Khan, A. H. & Pieroni, A. Ethnoveterinary plants of Pakistan: a review. *Journal of ethnobiology and ethnomedicine* vol. 16 25 Preprint at (2020). https://doi.org/10.1186/s13002-020-00369-1
66. Shuaib, M. et al. Traditional knowledge about medicinal plant in the remote areas of Wari Tehsil, Dir Upper, Pakistan. *Brazilian J. Biology*. **83**, 1–28 (2023).
67. Majid, A. et al. Exploring threatened traditional knowledge; ethnomedicinal studies of rare endemic flora from lesser himalayan region of Pakistan. *Revista Brasileira De Farmacognosia*. **29**, 785–792 (2019).
68. Hassan, N. et al. Identification and quantitative analyses of medicinal plants in Shahgram valley, district swat, Pakistan. *Acta Ecol. Sin.* **40**, 44–51 (2020).
69. Ashfaq, S. et al. Medicinal plant biodiversity used among the rural communities of arid regions of northern Punjab, Pakistan. *Indian J. Traditional Knowl.* **18**, 226–241 (2019).

70. Tali, B. A., Khuroo, A. A., Ganie, A. H. & Nawchoo, I. A. Diversity, distribution and traditional uses of medicinal plants in Jammu and Kashmir (J&K) state of Indian Himalayas. *J. Herb. Med.* **17–18**, 100280 (2019).
71. Guo, C. A. et al. An ethnobotany survey of wild plants used by the tibetan people of the Yadong River Valley, Tibet, China. *J. Ethnobiol. Ethnomed.* **18**, 1–25 (2022).
72. Figueiredo Junior, E. C. et al. Ethnobotanical knowledge of herbalists about medicinal plants from the semiarid region in northeastern Brazil. *Bol. Latinoam. Caribe Plantas Med. Aromat.* **21**, 803–815 (2022).
73. Belichenko, O., Kolosova, V., Kalle, R. & Sõukand, R. Green pharmacy at the tips of your toes: medicinal plants used by Setos and russians of Pechorsky District, Pskov Oblast (NW Russia). *J. Ethnobiol. Ethnomed.* **18**, 1–38 (2022).
74. Ahoyo, C. C. et al. Traditional medicinal knowledge of woody species across climatic zones in Benin (West Africa). *J. Ethnopharmacol.* **265** (2021).
75. Khajuria, A. K., Manhas, R. K., Kumar, H. & Bisht, N. S. Ethnobotanical study of traditionally used medicinal plants of Pauri district of Uttarakhand, India. *J. Ethnopharmacol.* **276**, 114204 (2021).
76. Lee, C. et al. Ethnobotanical study on medicinal plants used by local Van Kieu ethnic people of Bac Huong Hoa nature reserve, Vietnam. *J. Ethnopharmacol.* **231**, 283–294 (2019).
77. Al Hatmi, S. & Lupton, D. A. Documenting the most widely utilized plants and the potential threats facing ethnobotanical practices in the Western Hajar Mountains, sultanate of Oman. *J. Arid Environ.* **189**, 104484 (2021).

78. Caballero-Serrano, V. et al. Traditional ecological knowledge and medicinal plant diversity in Ecuadorian Amazon home gardens. *Glob Ecol. Conserv.* **17**, e00524 (2019).

79. Kimpouni, V., Mamboueni, J. C., Tsoungould, F. G. M. & Mikoko, E. N. Ethnobotanical and phytotherapeutic study from Kouni community of the sub-prefecture of Kayes (Bouenza – Congo). *Heliyon* **5**, e02007 (2019).

80. Cordoba-Tovar, L. et al. Cultural belief and medicinal plants in treating COVID 19 patients of western Colombia. *Acta Ecol. Sin.* https://doi.org/10.1016/j.chnaes.2021.10.011 (2021).

81. Fars Meteorological Organization. www.farsmet.ir/ReportAmar.aspx. (2020).

82. Sehati, M. T. Impact of slope and rock layers of conglomerate and sandstone lithology on vegetation cover in arid lands (Case study: Lamerd Fars Province). *Arid Biome Sci. Res. J.* **1**, 39–48 (2012).

83. Alalwan, T. A., Alkhuzai, J. A., Jameel, Z. & Mandeel, Q. A. Quantitative ethnobotanical study of some Medicinal plants used by herbalists in Bahrain. *J. Herb. Med.* **17–18**, 100278 (2019).

84. Moradi, G., Abbasi, E. & Zare Chahooki, M. A. Flora, life forms, and geographical distrbution of plants in Lamerd ranges, Fars, Iran. *Watershed management researchers (pajouhesh-va-sazandegi)* 86, 70–80 (2010).

85. Panmei, R., Gajurel, P. R. & Singh, B. Ethnobotany of medicinal plants used by the Zeliangrong ethnic group of Manipur, northeast India. *J. Ethnopharmacol.* **235**, 164–182 (2019).

86. Yebouk, C. et al. Ethnobotanical study of medicinal plants in the Adrar Province, Mauritania. *J. Ethnopharmacol.* **246**, 112217 (2020).

87. Miara, M. D. et al. Medicinal plants and their traditional uses in the highland region of Bordj Bou Arreridj (Northeast Algeria). *J. Herb. Med.* **16**, 100262 (2019).
88. Cochran, W. G. *Sampling Techniques* (Wiley Blackwell, 1977).
89. Espinosa, M. M., Bieski, I. G. & Martins, D. T. Sampling in ethnobotanical studies of medicinal plants. *Methods and Techniques in Ethnobiology and Ethnoecology* 197–212 (2014).
90. Tamang, S., Singh, A., Bussmann, R. W., Shukla, V. & Nautiyal, M. C. Ethno-medicinal plants of tribal people: a case study in Pakyong subdivision of East Sikkim, India. *Acta Ecol. Sin.* https://doi.org/10.1016/j.chnaes.2021.08.013 (2022).
91. Mozaffarian, V. *Dictionary of Iranian Plant Names: Latin-English-Persian* (Farhang moaser, 2009).
92. Mozaffarian, V. *Identification of Medicinal and Aromatic Plants of Iran* (Farhang moaser, 2016).
93. International Plant Names Index. www.ipni.org. (2022).
94. World Flora Online. www.worldfloraonline.org. (2022).
95. Plants of the World Online. www.plantsoftheworldonline.org. (2022).
96. The, I. U. C. N. & Red List. *www.iucnredlist.org* (2022).
97. International Classification Committee of WONCA. *International Classification of Primary Care* (Oxford University, 2005).
98. Hussain, S. et al. Quantitative ethnopharmacological profiling of medicinal shrubs used by indigenous communities of Rawalakot, District Poonch, Azad Jammu and Kashmir, Pakistan. *Revista Brasileira De Farmacognosia*. **29**, 665–676 (2019).

99. Sukumaran, S., Sujin, R. M., Geetha, V. S. & Jeeva, S. Ethnobotanical study of medicinal plants used by the Kani tribes of Pechiparai Hills, Western Ghats, India. *Acta Ecol. Sin.* **41**, 365–376 (2021).

100. Phillips, O., Gentry, A. H., Reynel, C., Wilkin, P. & Galvez-Durand, B. C. Etnobotánica cuantitativa y la conservación de la Amazonia. *Conserv. Biol.* **8**, 225–248 (1994).

101. Whitney, C. EthnobotanyR: Calculate quantitative ethnobotany indices. (2022). https://CRAN.R-project.org/package=ethnobotanyR

102. El-Ghazouani, F., El-Ouahmani, N., Teixidor-Toneu, I., Yacoubi, B. & Zekhnini, A. A survey of medicinal plants used in traditional medicine by women and herbalists from the city of Agadir, southwest of Morocco. *Eur. J. Integr. Med.* **42**, 101284 (2021).

103. Tardío, J. & Pardo-De-Santayana, M. Cultural importance indices: a comparative analysis based on the useful wild plants of southern Cantabria (northern Spain). *Econ. Bot.* **62**, 24–39 (2008).

104. Sargin, S. A. & Büyükcengiz, M. Plants Used in Ethnomedicinal Practices in Gulnar District of Mersin, Turkey. *J. Herb. Medi.* **15** (2019).

105. Polat, R. Ethnobotanical study on medicinal plants in Bingöl (City center) (Turkey). *J. Herb. Med.* **16** (2019).

106. Reyes-García, V., Huanca, T., Vadez, V., Leonard, W. & Wilkie, D. Cultural, practical, and economic value of wild plants: a quantitative study in the Bolivian Amazon. *Econ. Bot.* **60**, 62–74 (2006).

107. Kremen, C., Raymond, I. & Lance, K. An interdisciplinary tool for monitoring conservation impacts in Madagascar. *Conserv. Biol.* **12**, 549–563 (1998).

108. Chaachouay, N., Benkhnigue, O., Douira, A. & Zidane, L. Poisonous medicinal plants used in the popular pharmacopoeia of the Rif, northern Morocco. *Toxicon* **189**, 24–32 (2021).
109. Aparicio, H., Hedberg, I., Bandeira, S. & Ghorbani, A. Ethnobotanical study of medicinal and edible plants used in Nhamacoa area, Manica province–Mozambique. *South. Afr. J. Bot.* **139**, 318–328 (2021).
110. Nguyen, X. M. A., Bun, S. S., Ollivier, E. & Dang, T. P. T. Ethnobotanical study of medicinal plants used by K'Ho-Cil people for treatment of diarrhea in Lam Dong Province, Vietnam. *J. Herb. Med.* **19**, 100320 (2020).
111. Mosaddegh, M., Naghibi, F., Moazzeni, H., Pirani, A. & Esmaeili, S. Ethnobotanical survey of herbal remedies traditionally used in Kohghiluyeh Va Boyer Ahmad province of Iran. *J. Ethnopharmacol.* **141**, 80–95 (2012).
112. Saki, K. et al. The most common native medicinal plants used for psychiatric and neurological disorders in Urmia city, northwest of Iran. *Asian Pac. J. Trop. Dis.* **4**, S895–S901 (2014).
113. Savić, J., Mačukanović-Jocić, M. & Jarić, S. Medical ethnobotany on the Javor Mountain (Bosnia and Herzegovina). *Eur. J. Integr. Med.* **27**, 52–64 (2019).
114. Xiong, Y., Sui, X., Ahmed, S., Wang, Z. & Long, C. Ethnobotany and diversity of medicinal plants used by the Buyi in eastern Yunnan, China. *Plant. Divers.* **42**, 401–414 (2020).
115. Hosseini, S. H. & Bibak, H. Ramezani Ghra, A. Ethnobotanical study of medicinal plants of south Kerman. *Eco-phytochemical J. Med. Plants*. **1**, 30–63 (2020).

116. Bazgir, A. & Pour Hashemi, M. Ethnobotany of plants of the Bastam region of Selseleh county, Lorestan province. *Sci. J. Iran. Indigenous Knowl* **7**, 85–158 (2021).

117. Dolatkhahi, M., Ghorbani Nohooji, M., Mehrafarin, A., Amini Nejad, G. & Dolatkhahi, A. Ethnobotanical Study of Medicinal Plants in Kazeroon, Iran: identification, distribution and traditional usage. *J. Herb. Drugs* **2**, (2012).

118. Zolfaghari Baghersad, R., Piri, K., Abdoli, A., Mehrabian, A. R. & Abdoli, S. Importance of Using Ethnobiological Knowledge for the Conservation of Medicinal Plants Biodiversity in the Lar Region (Iran). *Journal of Medicinal Plants and By-products* (2012).

119. Razmjoue, D., Zarei, Z. & Armand R.

120. Malekpourzadeh, L., Ghahremaninejad, F. & Mirtajadini, S. M. Collection and Identification of Some Selected Medicinal Plants with Antimicrobial Properties from Takhte-Sartashtak Region, Kerman, Iran. *Journal of Medicinal Plants and By-products* (2012).

121. Mehrnia, M. & Hosseini, Z. Ethnobotanical study of medicinal plants native to Al-shatar region (Lorestan). *J. Traditional Med. Islam Iran.* **11**, 81–112 (2020).

122. Mir shekar, M., Ebrahimi, M. & Ajourloo, M. Ethnobotanical study and traditional uses of some medicinal plants of Khash County. *J. Traditional Med. Islam Iran.* **9**, 361–370 (2018).

123. Rahim foruzeh, M., Heshmati, G. & Barani, H. Collection and ethnobotanical study of selected plants from Kohgiluyeh and Boyer Ahmad province. *J. Traditional Med. Islam Iran.* **5**, 131–139 (2014).

124. Delfan, B., Bahmani, M., Hassanzadazar, H. & Saki, K. Ethnobotany study of effective medicinal plants on gastric problems in Lorestan Province, West of Iran. *J. Chem. Pharm. Res.* **7**, 483–492 (2015).

125. Khodayari, H., Amani, S. & Amiri, H. Ethnobotany of medicinal plants of northeastern Khuzestan Province. *J. Eco-phytochemistry Med. Plants*. **8**, 12–26 (2014).

126. Delfan, E., Khodayari, H. & Azizi, K. Ethnobotany of Native Medicinal Plants in Zagheh and Biranshahr districts, Lorestan Province, Iran. *Ech-Phytochem. J. Med. Plants* **4** (2019).

127. Parsaei, P., Bahmani, M., Naghdi, N. & Asadi-Samani, M. Identification of medicinal plants effective on common cold: an ethnobotanical study of Shiraz, South Iran. *Pharm. Lett.* **8**, 90–97 (2016).

128. Jivad, N., Bahmani, M. & Asadi-Samani, M. A review of the most important medicinal plants effective on wound healing on ethnobotany evidence of Iran. *Pharm. Lett.* **8**, 353–357 (2016).

129. Sharififar, F., Moharam Khani, M. R., Moattar, F., Babakhanloo, P. & Khodami, M. Ethnobotanical Study of Medicinal Plants of Joopar Mountains of Kerman Province, Iran. *J. Kerman Univ. Med. Sci.* **21**, 37–51 (2014).

130. Ahmadipour, S., Mohsenzadeh, A., Eftekhari, Z. & Ahmadipour, S. An overview of the most important medicinal plants affecting on child's jaundice in Ethnobotanical resource of Iran. *Pharm. Lett.* **8**, 135–139 (2016).

131. Mozaffari Nejad, A. S. & Laey, G. Ethnobotany and Folk Pharmaceutical properties of major trees or shrubs in Northeast of Iran. *Asian J. Chem.* **21**, 5632–5638 (2009).

132. Moradi, M. T., Asadi-Samani, M. & Bahmani, M. Hypotensive medicinal plants according to ethnobotanical evidence of Iran: a systematic review. *Int. J. PharmTech Rerearch*. **9**, 416–426 (2016).
133. Parsaei. Pouya, Bahmani, M., Asadi-Samani, M. & Rafieian-Kopaei, M. The most important medicinal plants effective on constipation by the ethnobotanical documents in Iran: a review. *Pharm. Lett.* **8**, 188–194 (2016).
134. Rouhi-Boroujeni, H., Majid, A. S. & Moradi, M. T. A review of the medicinal plants effective on headache based on the ethnobotanical documents of Iran. *Pharm. Lett.* **8**, 37–42 (2016).
135. Delfan, B. et al. Effective herbs on the wound and skin disorders: a ethnobotanical study in Lorestan province, west of Iran. *Asian Pac. J. Trop. Dis.* **4**, S938–S942 (2014).
136. Bahmani, M., Zargaran, A., Rafieian-Kopaei, M. & Saki, K. Ethnobotanical study of medicinal plants used in the management of diabetes mellitus in the Urmia, Northwest Iran. *Asian Pac. J. Trop. Med.* **7**, S348–S354 (2014).
137. Mosaddegh, M., Esmaeili, S., Hassanpour, A., Malekmohammadi, M. & Naghibi, F. Ethnobotanical Study in the Highland of Alvand and Tuyserkan, Iran. *Res. J. Pharmacognosy (RJP)* **3** (2016). http://rjpharmacognosy.ir
138. Ma'ameri, M., Abbasi khalaki, M. & Dadjoo, F. Ethnobotany of plants of Darvish Chai Sarein watershed with a medicinal and nutritional approach. *Sci. J. Rangel.* **14**, 698–714 (2020).
139. Moghanloo, L., Ghahremaninejad, F. & Vafadar, M. Ethnobotanical study of medicinal plants in the central district of the Zanjan County, Zanjan Province, Iran. *J. Herb. Med.* **9**, 121–131 (2019).

140. Saadatpour, M., Barani, H., Abedi Sarvestani, A. & Forouzeh, M. R. Ethnobotanical study of Sojasrood medicinal plants (Zanjan Province). *J. Herb. Drugs*. **8**, 185–193 (2017).
141. Mardani nejad S. & Vazir poor, M.
142. Abbasi, S., Afsharzadeh, S., Mohajeri, A. & reza Introduction of plant species with medicinal properties in Natanz region (Kashan). *J. Herb. Drugs*. **3**, 147–156 (2012).
143. Eskandari, M., Assadi, M., Shirzadian, S. & Mehregan, I. Ethnobotanical study and distribution of the Solanum section Solanum species (Solanaceae) in Iran. *J. Med. Plants*. **18**, 85–98 (2019).
144. Dolat khahi, M. Nabi poor, I. Ethnobotanical study of medicinal plants of the northeastern Persian Gulf watershed. *J. Herb. Drugs*. **2**, 129–143 (2014).
145. Ethnobotanical study of some medicinal plants. Of Abhar County, Zanjan Province. *J. Herb. Drugs*. **19**, 30–54 (2020).
146. Moradi, M. T., Asadi-Samani, M., Bahmani, M. & Shahrani, M. Medicinal plants used for liver disorders based on the ethnobotanical documents of Iran: a review. *Int. J. Pharmtech Res.* **9**, 407–415 (2016).
147. Mohamadi, N., Sharififar, F., Koohpayeh, A. & Daneshpajouh, M. Traditional and ethnobotanical uses of medicinal plants by ancient populations in Khabr and Rouchon of Iran. *J. Appl. Pharm. Sci.* **5**, 101–107 (2015).
148. Zahra Sadeghi, I. & Kuhestani, K. *Ethnobotany of Date Palm (Phoenix Dactylifera) in Baluch Tribe of Saravan Region. Journal of Agricultural Technology* vol. 10 www.naturland.de (2014).
149. Naghibi, F., Esmaeili, S., Malekmohammadi, M., Hassanpour, A. & Mosaddegh, M. Ethnobotanical survey of medicinal plants used traditionally

in two villages of Hamedan, Iran. *Res. J. Pharmacognosy (RJP)*. **1**, 7–14 (2014).

150. Kiyasi, Y. & Forouzeh, M. R. Ethnobotanical Study of Medicinal Plants in Abadeh County (Case Study: Almalicheh Range). *J. Tradit. Med. Islam Iran.* (2019).
151. Haerinasab, M. & Abbasi, S. Ethnobotanical study of medicinal plants and introduction to some poisonous plant species of Ardestan (Isfahan province). *J. Med. Plants*. **18**, 122–143 (2019).
152. Ramezanian, M. & Minaie far, A. Ethnobotanical Study of Medicinal plants of Fasa County. *J. Tradit. Med. Islam Iran.* **7** (2016).
153. Dastyar, N. & Altememy, D. Ethnobotanical Study for Hypertension Treatment: an ethnobotanical study of Abadeh, South Iran. *J. Biochemicals Phytomedicine*. **1**, 13–16 (2022).
154. Dastyar, N. & Lysiuk, R. Identification of the most important Medicinal plants used for Wound Healing: an ethnobotanical study of Sistan and Baluchestan Province, Southeastern Iran. *J. Biochemicals Phytomedicine*. **2**, 16–19 (2023).
155. Dastyar, N. & Ahmadi, M. An Ethnobotanical Study of Medicinal Plants administered for the Treatment of Hyperlipidemia in Bushehr, South Iran. *J. Biochemicals Phytomedicine*. **1**, 26–30 (2022).
156. Ghasemi Pirbalouti, A. Ethnobotany and antimicrobial activity of medicinal plants of Bakhtiari Zagross mountains, Iran. *J. Med. Plants Res.* **6** (2012).
157. Negahdari, S. Ethnobotanical Study of Medicinal Plants Used for Management of Diabetes Mellitus in the East of Khuzestan, Southwest Iran. *J. Biochemicals Phytomedicine*. **2**, 7–10 (2023).

158. Bahmani, M. *Overview of Medicinal plants used for Cardiovascular System disorders and diseases in Ethnobotany of different areas in Iran. J. HerbMed Pharmacol. J. Homepage: J. HerbMed Pharmacol.* **5** (2016). http://www.herbmedpharmacol.com
159. Nadaf, M. et al. Ethnobotanical Diversity of Trees and Shrubs of Iran: A Comprehensive Review. *International Journal of Plant Biology* vol. 14 120–146 Preprint at (2023). https://doi.org/10.3390/ijpb14010011
160. Hosseini, M., Rahim Forouzeh, M. & Barani, H. Identification and investigation of ethnobotany of some medicinal plants in Razavi Khorasan Province. *J. Med. Plants.* **18**, 212–231 (2019).
161. Mechaala, S., Bouatrous, Y. & Adouane, S. Traditional knowledge and diversity of wild medicinal plants in El Kantara's area (Algerian Sahara gate): an ethnobotany survey. *Acta Ecol. Sin.* **42**, 33–45 (2022).
162. Mogale, M. M. P., Raimondo, D. C. & VanWyk, B. E. The ethnobotany of Central Sekhukhuneland, South Africa. *South. Afr. J. Bot.* **122**, 90–119 (2019).
163. Alqethami, A., Aldhebiani, A. Y. & Teixidor-Toneu, I. Medicinal plants used in Jeddah, Saudi Arabia: a gender perspective. *J. Ethnopharmacol.* **257**, 112899 (2020).
164. Birjees, M. et al. Traditional knowledge of wild medicinal plants used by the inhabitants of Garam Chashma Valley, district Chitral, Pakistan. *Acta Ecol. Sin.* **42**, 19–33 (2022).
165. Ullah, A. et al. Ethnomedicinal flora of Frontier Region Tank, Fata, Pakistan. *Acta Ecol. Sin.* **39**, 321–327 (2019).
166. Lin, Y. et al. Ethnobotanical survey of medicinal plants in Gaomi, China. *J. Ethnopharmacol.* **265**, 113228 (2021).

PubMed Google Scholar Article

167. Al-Fatimi, M. Ethnobotanical survey of medicinal plants in central Abyan governorate, Yemen. *J. Ethnopharmacol.* **241**, 111973 (2019).
168. Prabhu, S., Vijayakumar, S., Morvin Yabesh, J. E., Prakashbabu, R. & Murugan, R. An ethnobotanical study of medicinal plants used in pachamalai hills of Tamil Nadu, India. *J. Herb. Med.* **25**, 100400 (2021).
169. Petrakou, K., Iatrou, G. & Lamari, F. N. Ethnopharmacological survey of medicinal plants traded in herbal markets in the peloponnisos, Greece. *J. Herb. Med.* **19**, 100305 (2020).
170. Abdusalam, A. et al. Documenting the heritage along the Silk Road: an ethnobotanical study of medicinal teas used in Southern Xinjiang, China. *J. Ethnopharmacol.* **260**, 113012 (2020).
171. Bhattarai, S., Chaudhary, R. P. & Taylor, R. S. L. Ethnomedicinal plants used by the people of Manang district, central Nepal. *J. Ethnobiol. Ethnomed.* **2**, 1–8 (2006).
172. Kyaw, Y. M. M., Bi, Y., Oo, T. N. & Yang, X. Traditional medicinal plants used by the Mon people in Myanmar. *J. Ethnopharmacol.* **265**, 113253 (2021).
173. Gumisiriza, H., Birungi, G., Olet, E. A. & Sesaazi, C. D. Medicinal plant species used by local communities around Queen Elizabeth National Park, Maramagambo Central Forest Reserve and Ihimbo Central Forest Reserve, South western Uganda. *J. Ethnopharmacol.* **239**, 111926 (2019).
174. Joshi, N., Ghorbani, A., Siwakoti, M. & Kehlenbeck, K. Utilization pattern and indigenous knowledge of wild medicinal plants among three ethnic groups in Makawanpur district, Central Nepal. *J. Ethnopharmacol.* **262**, 113219 (2020).

175. Roy, S. K. & Roy, D. K. Use of Medicinal Plant and its vulnerability due to Climate Change in Northern Part of Bangladesh. *Am. J. Plant. Sci.* **07**, 1782–1793 (2016).

176. Hounsou, K. E., Sonibare, M. A. & Elufioye, T. O. Climate change and the future of medicinal plants research. *Bioactive Compounds in Health and Disease* vol. 7 152–169 Preprint at (2024). https://doi.org/10.31989/BCHD.V7I3.1310

177. Stojanoski N. Development of health culture in Veles and its region from the past to the end of the 20th century. Veles: Society of science and art. 1999:13–34.

178. Kelly K. History of medicine. New York: Facts on file; 2009. pp. 29–50.

179. Bottcher H. Miracle drugs. Zagreb: Zora; 1965. pp. 23–139.

180. Wiart C. Etnopharmacology of medicinal plants. New Jersey: Humana Press; 2006. pp. 1–50.

181. Tucakov J. Healing with plants – phytotherapy. Beograd: Culture; 1971. pp. 180–90.

182. 6.Glesinger L. Medicine through centuries. Zagreb: Zora; 1954. pp. 21–38.

183. Tucakov J. Pharmacognosy. Beograd: Institute for text book issuing in SR. Srbije; 1964. pp. 11–30.

184. Dimitrova Z. The history of pharmacy. Sofija: St Clement of Ohrid; 1999. pp. 13–26.

185. Toplak Galle K. Domestic medicinal plants. Zagreb: Mozaic book; 2005. pp. 60–1.

186. Bojadzievski P. The health services in Bitola through the centuries. Bitola: Society of science and art; 1992. pp. 15–27.

187. Gorunovic M, Lukic P. Pharmacognosy. Beograd: Gorunovic M; 2001. pp. 1–5.
188. Pelagic V. Pelagic folk teacher. Beograd: Freedom; 1970. pp. 500–2.
189. Katic R. La medicine en Serbie au moyen age. Beograd: Scientific work; 1958. pp. 7–36.
190. Bazala V. The historical development of medicine in the Croatian lands. Zagreb: Croation publishing bibliographic institute; 1943. pp. 9–20.
191. Nikolovski B. Essays on the history of health culture in Macedonia. Skopje: Macedonian pharmaceutical association; 1995. pp. 17–27.
192. Tucakov J. Pharmacognosy. Beograd: Academic books; 1948. pp. 8–21.
193. Thorwald J. Power and knowledge of ancient physicians. Zagreb: August Cesarec; 1991. pp. 10–255.
194. Katic R. The Serbian medicine from 9th to 19th centuries. Beograd: Scientific work; 1967. pp. 22–37.
195. Stojcevska-Antic V. Clement and Naum of Ohrid in folk tradition. Skopje: Our book; 1982. pp. 25–86.
196. Celakoski N. Saint Naum of Ohrid Miracle worker. Prilep: Raster; 1997. pp. 85–6.
197. Nikolovski B. Arab pharmacy in Macedonia. Bulletin. 1961;1:20–7.
198. Katic R. In: In: The Chilandar medical codex N. 517. Milincevic V, editor. Beograd: National library from Srbija; 1980. pp. 9–80.
199. Tucakov J. Healing with plants. Beograd: Rad; 1990. pp. 576–8.
200. Tucakov J. Healing with plants. Beograd: Rad; 1990. pp. 24–37.
201. Jancic R. Botanika farmaceutika. Beograd: Public company Sl. List SRJ; 2002. pp. 83–6.

202. Dervendzi V. Contemporary treatment with medicinal plants. Skopje: Tabernakul; 1992. pp. 5–43.
203. Lukic P. Pharmacognosy. Beograd: SSO Faculty of Pharmacy; 1985. pp. 8–22.
204. Kovacevic N. Fundamentals of pharmacognosy. Beograd: Personal edition; 2000. pp. 170–1.
205. Nelson D, Cox M. Lehninger Principles of Biochemistry. 4th ed. New York: W.H. Freeman and Company; 2005. pp. 1–41.
206. Council of Europe, Strasburg. 6th ed 2008. European Pharmacopoeia.
207. The United States Pharmacopoeial Convention. Washington: 2008. USP 31 the United States Pharmacopoeia.
208. British Pharmacopoeia Commission. London: 2007. British Pharmacopoeia.
209. Blumenthal M. The Complete German Commission E Monographs, Special Expert Committee of the German Federal Institute for Drugs and Medical Devices. Austin: 1998.
210. Law of medicines and medical supplies, Official gazette of RM no.106/07.
211. P. Benetiere, Alimentation et carie dentaire. in ALIMENTATION PRECARITE Comité national d'hygiène et de Santé bucco-dentaire-Paris; Centre de Recherche et d'Information Nutritionnelles (CERIN) **22**, 11–13 (2003)http://www.cerin.org/periodiques/AlimPreca/AlimPreca22.asp. Accessed 15 Marsh 2017
212. Sheiham A. Oral health, general health and quality of life. Bull. World Health Organ. 2005;83:644.
213. Watt RG. Strategies and approaches in oral disease prevention and health promotion. Bull. World Health Organ. 2005;83(9):711–718.

214. Petersen PE. Global policy for improvement of oral health in the 21st century-implications to oral health research of World Health Assembly 2007, World Health Organization. Commun. Dent. Oral Epidemiol. 2009;37(1):1–8. doi: 10.1111/j.1600-0528.2008.00448.x.
215. Mak KK, Day JR. Dental health behaviours among early adolescents in Hong Kong. Int. J. Dent. Hyg. 2011;9(2):122–126. doi: 10.1111/j.1601-5037.2010.00452.x.
216. Petersen PE, Bourgeois D, Ogawa H, Estupinan-Day E, Ndiaye C. The global burden of oral diseases and risks to oral health. Bull. World Health Organ. 2005;83(9):661–669.
217. Palombo EA. Traditional medicinal plant extracts and natural products with activity against oral bacteria: potential application in the prevention and treatment of oral diseases. Evid Based Complement Alternat Med. 2011 doi: 10.1093/ecam/nep067.
218. World Health Organization, (2012) http://www.who.int/mediacentre/factsheets/fs318/fr/#. Accessed 13 December 2017
219. Zaoui F, Hamdani S, Belhad MJ, Miquel JL. Etude descriptive de l'état bucco-dentaire d'un échantillon de la population marocaine. Rev. Odont. Stomatol. Tropical. 1996;74:7–11.
220. Chukkapalli SS, Velsko IM, Rivera-Kweh MF, Zheng D, Lucas AR, Kesavalu L. Polymicrobial oral infection with four periodontal bacteria orchestrates a distinct inflammatory response and atherosclerosis in ApoE null mice. PLoS ONE. 2015;10(11):e0143291. doi: 10.1371/journal.pone.0143291.

221. J. Bellakhdar. Plantes médicinales au maghreb et soins de base-précis de phytothérapie moderne. (Eds Le Fennec, Casablanca, Maroc, 2006), pp. 386. ISBN: 9954-415-31-9

222. World Health Organization. Traditional Medicine Strategy 2002–2005. WHO. Geneva, 2002. Amsterdam, 2000

223. Kumar G, Jalaluddin M, Rout P, Mohanty R, Dileep CL. Emerging trends of herbal care in dentistry. J. Clin. Diagn Res. 2013;7:1827–1829. doi: 10.7860/JCDR/2013/6339.3282.

224. Taheri JB, Azimi S, Rafieiann N, Zanjani HA. Herbs in dentistry. Int. Dent. J. 2011;61:287–296. doi: 10.1111/j.1875-595X.2011.00064.x.

225. Sinha DJ, Sinha A. Natural medicaments in dentistry. Ayu. 2014;35:113–118. doi: 10.4103/0974-8520.146198.

226. Pellow T, Morsy M. The relation of Thomas Pellow: a reading of Morocco in the 18th century. Publ. Res. Civil. 1983;252:pp.

227. 17.High Commission for Planning. Monograph of the region of Casablanca. (2010)

228. Haut-commissariat au plan 1 (2013). http://www.hcp.ma/Recensement-general-de-la-population-et-de-l-habitat-2004_a633.html. Accessed 11 Dec 2017

229. 19.Haut-commissariat au plan 2 (2013). http://www.hcp.ma/Population-du-Maroc-par-annee-civile-en-milliers-et-au-milieu-de-l-annee-par-milieu-de-residence-1960-2050_a677.html. Accessed 11 Dec 2017

230. W. Greuter, H. M. Burder, G. Long, Med-checkliste. Med-checklist: a critical inventory of vascular plants of the circum-Mediterranean countries. Vol 1 (1984); vol. 3 (1986); vol. 4 (1989). Edit. Conservatoire et jardin Botaniques de Genere

231. T.G. Tutin, N.A. Burges, A.O. Chater, J.R. Edmondton, V.H. Heywood, D.M. Moore, D.H. Valentine, S.M. Walters, D.A. Webb (eds), Flora europaea 1, 2edn. Cambridge (1993)

232. The Plant List (The Plant List Version 1.1, 2013) http://www.theplantlist.org/. Accessed 1st April 2018

233. Phillips O, Gentry AH, Reynel C, Wilkin P, Gálvez Durand BC. Quantitative ethnobotany and Amazonian conservation. Conserv. Biol. 1994;8(1):225–248. doi: 10.1046/j.1523-1739.1994.08010225.x.

234. Vitalini S, Iriti M, Puricelli C, Ciuchi D, Segale A, Fico G. Traditional knowledge on medicinal and food plants used in Val San Giacomo (Sondrio, Italy)—an alpine ethnobotanical study. J. Ethnopharmacol. 2013;145(2):517–529. doi: 10.1016/j.jep.2012.11.024.

235. Suroowan S, Mahomoodally MF. A comparative ethnopharmacological analysis of traditional medicine used against respiratory tract diseases in Mauritius. J. Ethnopharmacol. 2016;177:61–80. doi: 10.1016/j.jep.2015.11.029.

236. Sreekeesoon DP, Mahomoodally MF. Ethnopharmacological analysis of medicinal plants and animals used in the treatment and management of pain in Mauritius. J. Ethnopharmacol. 2014;157:181–200. doi: 10.1016/j.jep.2014.09.030

237. Alexiades MN, Sheldon JW. Selected Guidelines for Ethnobotanical Research: A Field Manual. Boranx, NY: The New York Botanical Garden; 1996. pp. 99–133.

238. Fisseha M, Sebsebe D, Tilahun T. An ethnobotanical study of medicinal plants in WonagoWoreda, SNNPR, Ethiopia. J Ethnobiol Ethnomed. 2009;5(1):28. doi: 10.1186/1746-4269-5-28.

239. Gazzaneo LRS, De Lucena RFP, de Albuquerque UP. Knowledge and use of medicinal plants by local specialists in an region of Atlantic Forest in the state of Pernambuco (Northeastern Brazil) J. Ethnobiol. Ethnomed. 2005;1(1):9. doi: 10.1186/1746-4269-1-9.

240. Uddin MZ, Hassan MA. Determination of informant consensus factor of ethnomedicinal plants used in Kalenga forest, Bangladesh. Bangladesh J. Plant Taxon. 2014;21(1):83. doi: 10.3329/bjpt.v21i1.19272.

241. Howard P. The Major Importance of 'minor' Resources: Women and Plant Biodiversity. London: International Institute for Environment and Development (IIED); 2003.

242. B. Kothari, The invisible queen in the plant kingdom: Gender perspectives in medical ethnobotany. Women and Plants: Gender Relations in Biodiversity Management and Conservation. New York: Zed Books Ldt (2003): 150–164

243. Pfeiffer JM, Butz RJ. Assessing cultural and ecological variation in ethnobiological research: the importance of gender. J. Ethnobiol. 2005;25:240–278. doi: 10.2993/0278-0771(2005)25[240:ACAEVI]2.0.CO;2.

244. Elolemy AT, AlBedah AM. Public knowledge, attitude and practice of complementary and alternative medicine in Riyadh region, Saudi Arabia. Oman Med. J. 2012;27:20–26. doi: 10.5001/omj.2012.04.

245. Kankara SS, Ibrahim MH, Mustafa M, Go R. Ethnobotanical survey of medicinal plants used for traditional maternal healthcare in Katsina state, Nigeria. South African J. Bot. 2015;97:165–175. doi: 10.1016/j.sajb.2015.01.007.

246. Chonco WZ. The African Bantu traditional practice of medicine: some preliminary observations. Soc. Sci. Med. 1972;6(3):283–322. doi: 10.1016/0037-7856(72)90104-7.

247. Oladunmeye MK, Kehinde FY. Ethnobotanical survey of medicinal plants used in treating viral infections among Yoruba tribe South Western Nigeria. Afr. J. Microbiol. Res. 2011;5(19):2991–3004. doi: 10.4236/ajps.2014.521340.

248. Mehdioui R, Kahouadji A. Etude ethnobotanique auprès de la population riveraine de la forêt d'Amsittène: cas de la Commune d'Imi n'Tlit (Province d'Essaouira) Bulletin de l'Institut scientifique, Rabat, section Sciences de la vie. 2007;29:11–20.

249. Weniger B. Interest and limitation of a global ethnopharmacological survey. J. Ethnopharmacol. 1991;32:37–41. doi: 10.1016/0378-8741(91)90101-I.

250. Anyinam C. Ecology and ethnomedicine: exploring links between current environmental crisis and indigenous medical practices. Soc. Sci. Med. 1995;40(3):321–329. doi: 10.1016/0277-9536(94)E0098-D.

251. Benitez G, Gonzalez-Tejero MR, Molero-Mesa J. Pharmaceutical ethnobotany in the western part of Granada province (southern Spain): Ethnopharmacological synthesis. J. Ethnopharmacol. 2010;129:87–105. doi: 10.1016/j.jep.2010.02.016.

252. Savo V, Giulia C, Maria GP, David R. Folk phytotherapy of the Amalfi Coast (Campania, Southern Italy) J. Ethnopharmacol. 2011;135:376–392. doi: 10.1016/j.jep.2011.03.027.

253. Dey A, Gupta B, De JN. Traditional phytotherapy against skin diseases and in wound healing of the tribes of Purulia district, West Bengal, India. J. Med. Plants Res. 2012;6(33):4825–4831. doi: 10.5897/JMPR12.916.

254. Raja D, Blanché C, Vallès J. Contribution to the knowledge of the pharmaceutical ethnobotany of the Segarra region (Catalonia, Iberian Peninsula) J. Ethnopharmacol. 1997;57:149–160. doi: 10.1016/S0378-8741(97)00059-7.
255. Merzouki A, Ed-derfoufi F, Molero Mesa J. Contribution to the knowledge of Rifian traditional medicine. II: Folk medicine in Ksar Lakbir district (NW Morocco) Fitoterapia. 2000;71:278–307. doi: 10.1016/S0367-326X(00)00139-8.
256. Novais MH, Santos I, Mendes S, Pinto-Gomes C. Studies on pharmaceutical ethnobotany in Arrabida Natural Park (Portugal) J. Ethnopharmacol. 2004;93:183–195. doi: 10.1016/j.jep.2004.02.015.
257. Tahraoui A, El-Hilaly J, Israili ZH, Lyoussi B. Ethnopharmacological survey of plants used in the traditional treatment of hypertension and diabetes in south-eastern Morocco (Errachidia province) J. Ethnopharmacol. 2007;110:105–117. doi: 10.1016/j.jep.2006.09.011.
258. Ugulu I, Baslar S, Yorek N, Dogan Y. The investigation and quantitative ethnobotanical evaluation of medicinal plants used around Izmir province, Turkey. J. Med. Plants Res. 2009;3(5):345–367.
259. A. Sijelmassi. Les plantes médicinales du Maroc, 3ème édition Fennec. Casablanca, Moroc (1993)
260. Campanella L, Bonanni A, Favero G, et al. Determination of antioxidant properties of aromatic herbs, olives and fresh fruit using an enzymatic sensor. Anal. Bioanal. Chem. 2003;375:1011–1016. doi: 10.1007/s00216-003-1825-1.

261. Dragland S, Senoo H, Wake K. Several culinary and medicinal herbs are important sources of dietary antioxidants. J. Nut. 2003;133(5):1286–1290. doi: 10.1093/jn/133.5.1286.

262. Exarchou V, Nenadis N, Tsimidou M. Antioxidant activities and phenolic composition of extracts from greek oregano, greek sage, and summer savory. J. Agric. Food Chem. 2002;50:5294–5914. doi: 10.1021/jf020408a.

263. Lamiri A, Lhaloui S, Benjilali B. Insecticidal effects of essential oils against Hessian fly, Mayetiola destructor (Say) Field Crops Res. 2001;71(1):9–15. doi: 10.1016/S0378-4290(01)00139-3.

264. Cimanga K, Kambu K, Tona L, et al. Correlation between chemical composition and antibacterial activity of essential oils of some aromatic medicinal plants growing in the Democratic Republic of Congo. J. Ethnopharmacol. 2002;79(2):213–215. doi: 10.1016/S0378-8741(01)00384-1.

265. Lewis HA, Rudolph MJ, Mistry M, Monyatsi V, Marambana T, Ramela P. Oral health knowledge and original practices of African traditional healers in Zonkizizwe and Dube. South Africa. SADJ. 2004;59:243–246.

266. E.H.M. Ould, M. Hadj-Mahammed, H. Zabeirou. Place of spontaneous plants in the traditional medicine of the region of Ouargla (Northern Sahara), Courrier du Savoir. 03: 47–51 (2003). http://www.webreview.dz/IMG/pdf/6-Ould_H_didi.pdf

267. Benlamdini N, Elhafian M, Rochdi A, Zidane L. Etude floristique et ethnobotanique de la flore médicinale du haut Atlas oriental (Haute Moulouya) J. Appl. Biosci. 2014;78:6771–6787. doi: 10.4314/jab.v78i1.17.

268. Salhi S, Fadli M, Zidane L, Douira A. Etudes floristique et ethnobotanique des plantes médicinales de la ville de Kénitra (Maroc) Lazaroa. 2011;31:133–143. doi: 10.5209/rev_LAZA.2010.v31.9.

269. Muster D. Oral Medical Therapy: Means and Methods. Paris: Elsevier; 2004.

270. Pourabbas R, Delazar A, Chitsaz MT. The effect of German chamomile mouthwash on dental plaque and gingival inflammation. Iran. J. Pharmaceut. Res. 2005;2:105–109.

271. Pujar M, Makandar S. Herbal usage in endodontics: a review. Int. J. Contemp. Dent. 2011;2(1):34–37. doi: 10.1111/j.1365-2591.2009.01627.x.

272. Zhang Y, Wang Y, Zhu X, Cao P, Wei S, Lu Y. Antibacterial and antibiofilm activities of eugenol from essential oil of Syzygium aromaticum (L.) Merr. and L. M. Perry (clove) leaf against periodontal pathogen Porphyromonas gingivalis. Microb Pathog. 2017;113:396–402. doi: 10.1016/j.micpath.2017.10.054.

273. Sastravaha G, Yotnuengnit P, Booncong P, Sangtherapitikul P. Adjunctive periodontal treatment with Centella asiatica and Punica granatum extracts: a preliminary study. J. Int. Acad. Periodontol. 2003;5:106–115.

274. Sofrata A, Brito F, Al-Otaibi M, Gustafsson A. Short term clinical effect of active and inactive Salvadora persica miswak on dental plaque and gingivitis. J. Ethnopharmacol. 2011;137:1130–1134. doi: 10.1016/j.jep.2011.07.034.

275. Pai MB, Prashant GM, Murlikrishna KS, Shivakumar KM, Chandu GN. Antifungal efficacy of Punica granatum, Acacia nilotica, Cuminum cyminum and Foeniculum vulgare on Candida albicans: an in vitro study. Indian J. Dent. Res. 2010;21(3):334–336. doi: 10.4103/0970-9290.70792.

276. Dudhatra GB, Mody SK, Awale MM, Patel HB, Modi CM, Kumar A, Kamani DR, Chauhan BN. A Comprehensive review on pharmacotherapeutics of herbal bioenhancers. Sci. World J. 2012;2012:637953. doi: 10.1100/2012/637953.

277. Sharma P, Ravikumar G, Kalaiselvi M, Gomathi D, Uma C. In vitro antibacterial and free radical scavenging activity of green hull of Juglans regia. J. Pharm. Anal. 2013;3(4):298–302. doi: 10.1016/j.jpha.2013.01.006.

278. Zakavi F, Golpasand Hagh L, Daraeighadikolaei A, Farajzadeh Sheikh A, Daraeighadikolaei A, Leilavi Shooshtari Z. Antibacterial Effect of Juglans Regia Bark against Oral Pathologic Bacteria. Int. J. Dent. 2013;20:854765. doi: 10.1155/2013/854765.

279. Ghahraman A. Iranian Chromophytes. 1. Tehran: University Press Center; 1993. pp. 668–769.

280. M. Hmammouchi. Les Plantes Médicinales et Aromatiques Marocaines. Utilisations, Biologie, Ecologie, Chimie, Pharmacologie, Toxicologie et Lexiques Ed. Imprimerie Fédala. Morocco: Rabat- Instituts, p. 450 (1999)

281. J. Valnet. La phytothérapie-Traitement des maladies par les plantes-Se soigner par les plantes. VIGOT. ISBN. 2-253(2001)

282. Yalya EM, Izgu N, Ozdemir L, Erdem SA, Kartal SA. Sage tea–thyme–peppermint hydrosol oral rinse reduces chemotherapy-induced oral mucositis: a randomized controlled pilot study. Complement. Therapies in Med. 2016;27:58–64. doi: 10.1016/j.ctim.2016.05.010.

283. Khan ST, Khan M, Ahmad J, Wahab R, Abd-Elkader OH, Musarrat J, et al. Thymol and carvacrol induce autolysis, stress, growth inhibitionand reduce the biofilm formation by Streptococcus mutans. AMB Express. 2017;7:49. doi: 10.1186/s13568-017-0344-y.

284. Srithi K, Balslev H, Wangpakapattanawong P, Srisanga P, Trisonthi C. Medicinal plant knowledge and its erosion among the Mien (Yao) in northern Thailand. J. Ethnopharmacol. 2009;123:335–342. doi: 10.1016/j.jep.2009.02.035.
285. Bibi T, Ahmad M, Tareen RB, Tareen NM, Jabeen R, Rehman S. Ethnobotany of medicinal plants in district Mastung of Balochistan province-Pakistan. J. Ethnopharmacol. 2014;157:79–89. doi: 10.1016/j.jep.2014.08.042.
286. Benkhnigue O, Zidane L, Fadli M, Elyacoubi H, Rochdi A, Douira A. Ethnobotanical study of medicinal plants in the region of Mechraâ Bel Ksiri (Gharb region of Morocco) Acta Bot. Barc. 2011;53:191–216.
287. Taylor L. The Healing Power of Rainforest Herbs: A Guide to Understanding and Using Herbal Medicinals. New York: Square One Publishers; 2005.
288. Kathmandu RY. The burden of restorative dental treatment for children in Third World countries. Int. Dent. J. 2002;52(1):1–9. doi: 10.1111/j.1875-595X.2002.tb00589.x.
289. Cunninghan AB. African Medicinal Plants. Paris: United Nations Educational, Scientific and Cultural Organization; 1993.
290. Kasuku W, Lula F, Paulus J, Ngiefu N, Kaluila D. Contribution to the inventory of plants used for the treatment of malaria in Kinshasa (R.D.C.) J. Med. Afr. Pharmacop. 1999;13:95–103.
291. Dounias E, Rodrigue W, Petit C. Revue de la littérature ethnobotanique pour l'Afrique Centrale et l'Afrique de l'Ouest = Review of ethnobotanical literature for Central and West Africa.". Bulletin du Réseau Africain d'Ethnobotanique. 2000;2:5–117.

292. Alqareer A, Alyahya A, Andersson L. The effect of clove and benzocaine versus placebo as topical anesthetics. J. Dent. 2006;34(10):747–750. doi: 10.1016/j.jdent.2006.01.009.

293. Gomiero T. Soil degradation, land scarcity and food security: Reviewing a complex challenge. Sustainability. 2016;8:281. doi: 10.3390/su8030281.

294. Nambiar K.K.M., Gupta A.P., Fu Q.L., Li S. Biophysical, chemical and socio-economic indicators for assessing agricultural sustainability in the Chinese coastal zone. Agric. Ecosyst. Environ. 2001;87:209–214. doi: 10.1016/S0167-8809(01)00279-1.

295. Trivedi P., Delgado-Baquerizo M., Anderson I.C., Singh B.K. Response of soil properties and microbial communities to agriculture: Implications for primary productivity and soil health indicators. Front. Plant Sci. 2016;7:990. doi: 10.3389/fpls.2016.00990.

296. Turmel M.S., Speratti A., Baudron F., Verhulst N., Govaerts B. Crop residue management and soil health: A systems analysis. Agric. Syst. 2015;134:6–16. doi: 10.1016/j.agsy.2014.05.009.

297. Congreves K.A., Hayes A., Verhallen E.A., Van Eerd L.L. Long-term impact of tillage and crop rotation on soil health at four temperate agroecosystems. Soil Tillage Res. 2015;152:17–28. doi: 10.1016/j.still.2015.03.012.

298. Doran J.W., Zeiss M.R. Soil health and sustainability: Managing the biotic component of soil quality. Appl. Soil Ecol. 2000;15:3–11. doi: 10.1016/S0929-1393(00)00067-6.

299. De Corato U. Disease-suppressive compost enhances natural soil suppressiveness against soil-borne plant pathogens: A critical review. Rhizosphere. 2020;13:100192. doi: 10.1016/j.rhisph.2020.100192.

300. van Bruggen A.H.C., Semenov A.M. In search of biological indicators for soil health and disease suppression. Appl. Soil Ecol. 2000;15:13–24. doi: 10.1016/S0929-1393(00)00068-8.

301. De Corato U. Soil microbiota manipulation and its role in suppressing soil-borne plant pathogens in organic farming systems under the light of microbiome-assisted strategies. Chem. Biol. Technol. Agric. 2020;7:17. doi: 10.1186/s40538-020-00183-7.

302. Hannula S.E., Ma H., Pérez-Jaramillo J.E., Pineda A., Bezemer T.M. Structure and ecological function of the soil microbiome affecting plant–soil feedbacks in the presence of a soil-borne pathogen. Environ. Microbiol. 2020;22:660–676. doi: 10.1111/1462-2920.14882.

303. Mendes L.W., Mendes R., Raaijmakers J.M., Tsai S.M. Breeding for soil-borne pathogen resistance impacts active rhizosphere microbiome of common bean. ISME J. 2018;12:3038–3042. doi: 10.1038/s41396-018-0234-6.

304. Wolfgang A., Taffner J., Guimarães R.A., Coyne D., Berg G. Novel strategies for soil-borne diseases: Exploiting the microbiome and volatile-based mechanisms toward controlling Meloidogyne-based disease complexes. Front. Microbiol. 2019;10:1296. doi: 10.3389/fmicb.2019.01296.

305. Haas D., Défago G. Biological control of soil-borne pathogens by fluorescent pseudomonads. Nat. Rev. Microbiol. 2005;3:307–319. doi: 10.1038/nrmicro1129.

306. Hernández-Lambraño R.E., González-Moreno P., Sánchez-Agudo J.Á. Environmental factors associated with the spatial distribution of invasive plant pathogens in the Iberian Peninsula: The case of Phytophthora

cinnamomi Rands. For. Ecol. Manag. 2018;419–420:101–109. doi: 10.1016/j.foreco.2018.03.026.

307. Katan J. Diseases caused by soilborne pathogens: Biology, management and challenges. J. Plant Pathol. 2017;99:305–315. doi: 10.4454/jpp.v99i2.3862.
308. Mehta C.M., Palni U., Franke-Whittle I.H., Sharma A.K. Compost: Its role, mechanism and impact on reducing soil-borne plant diseases. Waste Manag. 2014;34:607–622. doi: 10.1016/j.wasman.2013.11.012.
309. Li X.G., Zhang T.L., Wang X.X., Hua K., Zhao L., Han Z.M. The composition of root exudates from two different resistant peanut cultivars and their effects on the growth of soil-borne pathogen. Int. J. Biol. Sci. 2013;9:164–173. doi: 10.7150/ijbs.5579. [DOI] [PMC free article] [PubMed] [Google Scholar]
310. van Elsas J.D., Garbeva P., Salles J. Effects of agronomical measures on the microbial diversity of soils as related to the suppression of soil-borne plant pathogens. Biodegradation. 2002;13:29–40. doi: 10.1023/A:1016393915414.
311. Tang J., Xue Z.Q., Daroch M., Ma J. Impact of continuous Salvia miltiorrhiza cropping on rhizosphere actinomycetes and fungi communities. Ann. Microbiol. 2015;65:1267–1275. doi: 10.1007/s13213-014-0964-2.
312. Bailey K.L., Lazarovits G. Suppressing soil-borne diseases with residue management and organic amendments. Soil Tillage Res. 2003;72:169–180. doi: 10.1016/S0167-1987(03)00086-2.
313. Du F.Y., Ju G.L., Xiao L., Zhou Y.M., Wu X. Sesquiterpenes and cyclodepsipeptides from marine-derived fungus Trichoderma longibrachiatum and their antagonistic activities against soil-borne pathogens. Mar. Drugs. 2020;18:165. doi: 10.3390/md18030165.

314. Panth M., Hassler S.C., Baysal-Gurel F. Methods for management of soilborne diseases in crop production. Agriculture. 2020;10:16. doi: 10.3390/agriculture10010016.
315. Barua A., McDonald-Howard K.L., Mc Donnell R.J., Rae R., Williams C.D. Toxicity of essential oils to slug parasitic and entomopathogenic nematodes. J. Pest Sci. 2020;93:1411–1419. doi: 10.1007/s10340-020-01251-5.
316. Cheng H.Y., Zhang D.Q., Ren L.R., Song Z.X., Li Q.J., Wu J.J., Fang W.S., Huang B., Yan D.D., Li Y., et al. Bio-activation of soil with beneficial microbes after soil fumigation reduces soil-borne pathogens and increases tomato yield. Environ. Pollut. 2021;283:117160. doi: 10.1016/j.envpol.2021.117160.
317. Stuart A.K.C., Stuart R.M., Pimentel I.C. Effect of agrochemicals on endophytic fungi community associated with crops of organic and conventional soybean (Glycine max L. Merril) Agric. Nat. Resour. 2018;52:388–392. doi: 10.1016/j.anres.2018.10.005
318. Ashraf S., Zuhaib M. Fungal biodiversity: A potential tool in plant disease management. In: Malik A., Grohmann E., Alves M., editors. Management of Microbial Resources in the Environment. Springer; Dordrecht, The Netherlands: 2013. pp. 69–90.
319. Verchot-Lubicz J. Soilborne viruses: Advances in virus movement, virus induced gene silencing, and engineered resistance. Physiol. Mol. Plant Pathol. 2003;62:55–63. doi: 10.1016/S0885-5765(03)00040-7.
320. Colla P., Gilardi G., Gullino M.L. A review and critical analysis of the European situation of soilborne disease management in the vegetable sector. Phytoparasitica. 2012;40:515–523. doi: 10.1007/s12600-012-0252-2.

321. Lougraimzi H., El Iraqui S., Bouaichi A., Gouit S., El Hassan A., Fadli M. Insecticidal effect of essential oil and powder of Mentha pulegium L. leaves against Sitophilus oryzae (Linnaeus, 1763) and Tribolium castaneum (Herbst, 1797) (Coleoptera: Curculionidae, Tenebrionidae), the main pests of stored wheat in Morocco. Pol. J. Entomol. 2018;87:263–278. doi: 10.2478/pjen-2018-0018.

322. Thampi A., Bhai R.S. Rhizosphere actinobacteria for combating Phytophthora capsici and Sclerotium rolfsii, the major soil borne pathogens of black pepper (Piper nigrum L.) Biol. Control. 2017;109:1–13. doi: 10.1016/j.biocontrol.2017.03.006.

323. Farzana, Abid M., Hussain F. Screening of ethnomedicinal plants for their antifungal and nematicidal activities against soil-borne phytopathogens. S. Afr. J. Bot. 2022;147:18–23. doi: 10.1016/j.sajb.2021.12.003.

324. Govindarajan M., Kadaikunnan S., Alharbi N.S., Benelli G. Acute toxicity and repellent activity of the Origanum scabrum Boiss. & Heldr. (Lamiaceae) essential oil against four mosquito vectors of public health importance and its biosafety on non-target aquatic organisms. Environ. Sci. Pollut. Res. 2016;23:23228–23238. doi: 10.1007/s11356-016-7568-2.

325. Kadoglidou K., Chatzopoulou P., Maloupa E., Kalaitzidis A., Ghoghoberidze S., Katsantonis D. Mentha and oregano soil amendment induces enhancement of tomato tolerance against soilborne diseases, yield and quality. Agronomy. 2020;10:406. doi: 10.3390/agronomy10030406.

326. Pretorius J.C., Zietsman P.C., Eksteen D. Fungitoxic properties of selected South African plant species against plant pathogens of economic importance in agriculture. Ann. Appl. Biol. 2002;141:117–124. doi: 10.1111/j.1744-7348.2002.tb00203.x.

327. Amoabeng B.W., Johnson A.C., Gurr G.M. Natural enemy enhancement and botanical insecticide source: A review of dual use companion plants. Appl. Entomol. Zool. 2019;54:1–19. doi: 10.1007/s13355-018-00602-0.
328. Shafique H.A., Athar M., Ehteshamul-Haque S., Sultana V. Management of soil-borne diseases of organic vegetables. J. Plant Prot. Res. 2016;56:221–230. doi: 10.1515/jppr-2016-0043.
329. Tahat M.M., Alananbeh K.M., Othman Y.A., Leskovar D.I. Soil health and sustainable agriculture. Sustainability. 2020;12:4859. doi: 10.3390/su12124859.
330. Bonanomi G., De Filippis F., Cesarano G., La Storia A., Ercolini D., Scala F. Organic farming induces changes in soil microbiota that affect agro-ecosystem functions. Soil Biol. Biochem. 2016;103:327–336. doi: 10.1016/j.soilbio.2016.09.005.
331. Shokoohi E., Mashela P.W., Machado R.A.R. Bacterial communities associated with Zeldia punctata, a bacterivorous soil-borne nematode. Int. Microbiol. 2022;25:207–216. doi: 10.1007/s10123-021-00207-8.
332. Bakker P.A.H.M., Berendsen R.L., Van Pelt J.A., Vismans G., Yu K., Li E., Van Bentum S., Poppeliers S.W.M., Gil J.J.S., Zhang H., et al. The soil-borne identity and microbiome-assisted agriculture: Looking back to the future. Mol. Plant. 2020;13:1394–1401. doi: 10.1016/j.molp.2020.09.017.
333. Zhao J., Liu J., Liang H., Huang J., Chen Z., Nie Y.J., Wang C.B., Wang Y.G. Manipulation of the rhizosphere microbial community through application of a new bio-organic fertilizer improves watermelon quality and health. PLoS ONE. 2018;13:e0192967. doi: 10.1371/journal.pone.0192967.
334. Sarwar M., Aslam M., Sarwar S., Iftikhar R. Different nematodes and plasmodiophorids as vectors of plant viruses. In: Awasthi L.P., editor.

Applied Plant Virology: Advances, Detection, and Antiviral Strategies. Academic Press; London, UK: 2020. pp. 275–290.

335. van Ruijven J., Ampt E., Francioli D., Mommer L. Do soil-borne fungal pathogens mediate plant diversity–productivity relationships? Evidence and future opportunities. J. Ecol. 2020;108:1810–1821. doi: 10.1111/1365-2745.13388.

336. Lucas P. Diseases caused by soil-borne pathogens. In: Cooke B.M., Jones D.G., Kaye B., editors. The Epidemiology of Plant Diseases. Springer; Dordrecht, The Netherlands: 2006. pp. 373–386.

337. Delgado-Baquerizo M., Guerra C.A., Cano-Díaz C., Egidi E., Wang J.T., Eisenhauer N., Singh B.K., Maestre F.T. The proportion of soil-borne pathogens increases with warming at the global scale. Nat. Clim. Chang. 2020;10:550–554. doi: 10.1038/s41558-020-0759-3.

338. Moutassem D., Belabid L., Bellik Y., Ziouche S., Baali F. Efficacy of essential oils of various aromatic plants in the biocontrol of Fusarium wilt and inducing systemic resistance in chickpea seedlings. Plant Prot. Sci. 2019;55:202–217. doi: 10.17221/134/2018-PPS.

339. Arora H., Sharma A., Poczai P., Sharma S., Haron F.F., Gafur A., Sayyed R.Z. Plant-derived protectants in combating soil-borne fungal infections in tomato and chilli. J. Fungi. 2022;8:213. doi: 10.3390/jof8020213.

340. Baysal-Gurel F., Kabir M.N., Liyanapathiranage P. Effect of organic inputs and solarization for the suppression of Rhizoctonia solani in woody ornamental plant production. Plants. 2019;8:138. doi: 10.3390/plants8050138.

341. Wang M., Ling N., Dong X., Zhu Y.Y., Shen Q.R., Guo S.W. Thermographic visualization of leaf response in cucumber plants infected

with the soil-borne pathogen Fusarium oxysporum f. sp. cucumerinum. Plant Physiol. Biochem. 2012;61:153–161. doi: 10.1016/j.plaphy.2012.09.015.

342. Malandrakis A.A., Kavroulakis N., Chrysikopoulos C.V. Use of copper, silver and zinc nanoparticles against foliar and soil-borne plant pathogens. Sci. Total Environ. 2019;670:292–299. doi: 10.1016/j.scitotenv.2019.03.210.

343. Erdoğan O., Çelik A., Zeybek A. In vitro antifungal activity of mint, thyme, lavender extracts and essential oils on Verticillium dahliae Kleb. Fresenius Environ. Bull. 2016;25:4856–4862.

344. Ayala-Doñas A., de Cara-García M., Talavera-Rubia M., Verdejo-Lucas S. Management of soil-borne fungi and root-knot nematodes in cucurbits through breeding for resistance and grafting. Agronomy. 2020;10:1641. doi: 10.3390/agronomy10111641.

345. Farhang V., Amini J., Ebadollahi A., Sadeghi G.R. Ocimum basilicum L. essential oil cultivated in Iran: Chemical composition and antifungal activity against three Phytophthora species. Arch. Phytopathol. Plant Prot. 2014;47:1696–1703. doi: 10.1080/03235408.2013.855022.

346. Mansfield J., Genin S., Magori S., Citovsky V., Sriariyanum M., Ronald P., Dow M., Verdier V., Beer S.V., Machado M.A., et al. Top 10 plant pathogenic bacteria in molecular plant pathology. Mol. Plant Pathol. 2012;13:614–629. doi: 10.1111/j.1364-3703.2012.00804.x.

347. Xue H., Lozano-Durán R., Macho A.P. Insights into the root invasion by the plant pathogenic bacterium Ralstonia solanacearum. Plants. 2020;9:516. doi: 10.3390/plants9040516.

348. Djebaili R., Pellegrini M., Ercole C., Farda B., Kitouni M., Del Gallo M. Biocontrol of soil-borne pathogens of Solanum lycopersicum L. and Daucus

carota L. by plant growth-promoting actinomycetes: In vitro and in planta antagonistic activity. Pathogens. 2021;10:1305. doi: 10.3390/pathogens10101305.
349. de Haan E.G., Dekker-Nooren T.C.E.M., van den Bovenkamp G.W., Speksnijder A.G.C.L., van der Zouwen P.S., van der Wolf J.M. Pectobacterium carotovorum subsp. carotovorum can cause potato blackleg in temperate climates. Eur. J. Plant Pathol. 2008;122:561–569. doi: 10.1007/s10658-008-9325-y.
350. Mwangi M., Bandyopadhyay R., Ragama P., Tushemereirwe W.K. Assessment of banana planting practices and cultivar tolerance in relation to management of soilborne Xanthomonas campestris pv musacearum. Crop Prot. 2007;26:1203–1208. doi: 10.1016/j.cropro.2006.10.017.
351. Kantor M., Handoo Z., Kantor C., Carta L. Top ten most important U.S.-regulated and emerging plant-parasitic nematodes. Horticulturae.
352. Collange B., Navarrete M., Montfort F., Mateille T., Tavoillot J., Martiny B., Tchamitchian M. Alternative cropping systems can have contrasting effects on various soil-borne diseases: Relevance of a systemic analysis in vegetable cropping systems. Crop Prot. 2014;55:7–15. doi: 10.1016/j.cropro.2013.10.002.
353. Jones J.T., Haegeman A., Danchin E.G.J., Gaur H.S., Helder J., Jones M.G.K., Kikuchi T., Manzanilla-López R., Palomares-Rius J.E., Wesemael W.M.L., et al. Top 10 plant-parasitic nematodes in molecular plant pathology. Mol. Plant Pathol. 2013;14:946–961. doi: 10.1111/mpp.12057.
354. Liu T.T., Wu H.B., Jiang H.Y., Zhang L., Zhang Y.N., Mao L.G. Thiophenes from Echinops grijsii as a preliminary approach to control disease complex of root-knot nematodes and soil-borne fungi: Isolation,

activities, and structure–nonphototoxic activity relationship analysis. J. Agric. Food Chem. 2019;67:6160–6168. doi: 10.1021/acs.jafc.9b01306.

355. Nguyen V.C., Villate L., Gutierrez-Gutierrez C., Castillo P., Van Ghelder C., Plantard O., Esmenjaud D. Phylogeography of the soil-borne vector nematode Xiphinema index highly suggests Eastern origin and dissemination with domesticated grapevine. Sci. Rep. 2019;9:7313. doi: 10.1038/s41598-019-43812-4.

356. Ntalli N.G., Ozalexandridou E.X., Kasiotis K.M., Samara M., Golfinopoulos S.K. Nematicidal activity and phytochemistry of greek Lamiaceae species. Agronomy. 2020;10:1119. doi: 10.3390/agronomy10081119.

357. Ntalli N.G., Ferrari F., Giannakou I., Menkissoglu-Spiroudi U. Phytochemistry and nematicidal activity of the essential oils from 8 Greek Lamiaceae aromatic plants and 13 terpene components. J. Agric. Food Chem. 2010;58:7856–7863. doi: 10.1021/jf100797m.

358. Vieira P., Gleason C. Plant-parasitic nematode effectors—Insights into their diversity and new tools for their identification. Curr. Opin. Plant Biol. 2019;50:37–43. doi: 10.1016/j.pbi.2019.02.007.

359. Palomares-Rius J.E., Castillo P., Navas-Cortés J.A., Jiménez-Díaz R.M., Tena M. A proteomic study of in-root interactions between chickpea pathogens: The root-knot nematode Meloidogyne artiellia and the soil-borne fungus Fusarium oxysporum f. sp. ciceris race 5. J. Proteomics. 2011;74:2034–2051. doi: 10.1016/j.jprot.2011.05.026.

360. Chen L., Daub M., Luigs H.G., Jansen M., Strauch M., Merhof D. High-throughput phenotyping of nematode cysts. Front. Plant Sci. 2022;13:965254. doi: 10.3389/fpls.2022.965254.

361. Khan M.R. Nematode infestation, a potential threat to Indian forests. Indian Phytopathol. 2020;73:397–414. doi: 10.1007/s42360-020-00228-9.

362. Ankrom K.E., Franco A.L.C., Fonte S.J., Gherardi L.A., de Tomasel C.M., Andriuzzi W.S., Shaw E.A., Sala O.E., Wall D.H. Ecto- and endoparasitic nematodes respond differently across sites to changes in precipitation. Oecologia. 2020;193:761–771. doi: 10.1007/s00442-020-04708-7.

363. Sorribas F.J., Djian-Caporalino C., Mateille T. Nematodes. In: Gullino M.L., Albajes R., Nicot P.C., editors. Integrated Pest and Disease Management in Greenhouse Crops. Springer; Cham, Switzerland: 2020. pp. 147–174.

364. Abad P., Gouzy J., Aury J.M., Castagnone-Sereno P., Danchin E.G.J., Deleury E., Perfus-Barbeoch L., Anthouard V., Artiguenave F., Blok V.C., et al. Genome sequence of the metazoan plant-parasitic nematode Meloidogyne incognita. Nat. Biotechnol. 2008;26:909–915. doi: 10.1038/nbt.1482.

365. Dong L.Q., Zhang K.Q. Microbial control of plant-parasitic nematodes: A five-party interaction. Plant Soil. 2006;288:31–45. doi: 10.1007/s11104-006-9009-3.

366. Wu H.M., Yan W.K., Wu H.M., Zhang J.Y., Zhang Z., Zhang Z.Y., Rensing C., Lin W.X. Consecutive monoculture regimes differently affected the diversity of the rhizosphere soil viral community and accumulated soil-borne plant viruses. Agric. Ecosyst. Environ. 2022;337:108076. doi: 10.1016/j.agee.2022.108076.

367. Waziri H.M. Plants as antiviral agents. J. Plant Pathol. Microbiol. 2015;6:1000254. doi: 10.4172/2157-7471.1000254.

368. Laufer M., Mohammad H., Christ D.S., Riedel D., Maiss E., Varrelmann M., Liebe S. Fluorescent labelling of Beet necrotic yellow vein virus and Beet

soil-borne mosaic virus for co- and superinfection experiments in Nicotiana benthamiana. J. Gen. Virol. 2018;99:1321–1330. doi: 10.1099/jgv.0.001122.
369. Abdelkhalek A., Salem M.Z.M., Hafez E., Behiry S.I., Qari S.H. The phytochemical, antifungal, and first report of the antiviral properties of Egyptian Haplophyllum tuberculatum extract. Biology. 2020;9:248. doi: 10.3390/biology9090248.
370. Triolo E., Materazzi A. Rate of tobacco mosaic virus degradation in solarized soil. [(accessed on 17 January 2023)];Riv. Patol. Veg. 1992 2:23–32. Available online: https://www.jstor.org/stable/42568064
371. Yoon J.B., Choi S.K., Cho I.S., Kwon T.R., Yang C.Y., Seo M.H., Yoon J.Y. Epidemiology of tomato spotted wilt virus in Chrysanthemum morifolium in South Korea and its management using a soil-dwelling predatory mite (Stratiolaelaps scimitus) and essential oils. Virus Res. 2020;289:198128. doi: 10.1016/j.virusres.2020.198128.
372. Taglienti A., Donati L., Ferretti L., Tomassoli L., Sapienza F., Sabatino M., Di Massimo G., Fiorentino S., Vecchiarelli V., Nota P., et al. In vivo antiphytoviral activity of essential oils and hydrosols from Origanum vulgare, Thymus vulgaris, and Rosmarinus officinalis to control zucchini yellow mosaic virus and tomato leaf curl New Delhi virus in Cucurbita pepo L. Front. Microbiol. 2022;13:840893. doi: 10.3389/fmicb.2022.840893.
373. Gakuya D.W., Itonga S.M., Mbaria J.M., Muthee J.K., Musau J.K. Ethnobotanical survey of biopesticides and other medicinal plants traditionally used in Meru central district of Kenya. J. Ethnopharmacol. 2013;145:547–553. doi: 10.1016/j.jep.2012.11.028.
374. Vlassi A., Nesler A., Perazzolli M., Lazazzara V., Büschl C., Parich A., Puopolo G., Schuhmacher R. Volatile organic compounds from Lysobacter

capsici AZ78 as potential candidates for biological control of soilborne plant pathogens. Front. Microbiol. 2020;11:1748. doi: 10.3389/fmicb.2020.01748.
375. Ajwa H.A., Klose S., Nelson S.D., Minuto A., Gullino M.L. Alternatives to methyl bromide in strawberry production in the United States of America and the Mediterranean region. Phytopathol. Mediterr. 2003;42:220–244. doi: 10.14601/Phytopathol_Mediterr-1715.
376. Mongiano G., Zampieri E., Morcia C., Titone P., Volante A., Terzi V., Tamborini L., Valé G., Monaco S. Application of plant-derived bioactive compounds as seed treatments to manage the rice pathogen Fusarium fujikuroi. Crop Prot. 2021;148:105739. doi: 10.1016/j.cropro.2021.105739.
377. Xie H.W., Yan D.D., Mao L.G., Wang Q.X., Li Y., Ouyang C.B., Guo M.X., Cao A.C. Evaluation of methyl bromide alternatives efficacy against soil-borne pathogens, nematodes and soil microbial community. PLoS ONE. 2015;10:e0117980. doi: 10.1371/journal.pone.0117980.
378. Verjux N. Integrated plant protection in potatoes: Review of the French situation. Potato Res. 2017;60:307–318. doi: 10.1007/s11540-018-9382-3.
379. Rodrigues G.C.S., dos Santos Maia M., Cavalcanti A.B.S., de Sousa N.F., Scotti M.T., Scotti L. In silico studies of Lamiaceae diterpenes with bioinsecticide potential against Aphis gossypii and Drosophila melanogaster. Molecules. 2021;26:766. doi: 10.3390/molecules26030766.
380. Arnault I., Fleurance C., Vey F., Fretay G.D., Auger J. Use of Alliaceae residues to control soil-borne pathogens. Ind. Crops Prod. 2013;49:265–272. doi: 10.1016/j.indcrop.2013.05.007.
381. Mao L.G., Jiang H.Y., Zhang L., Zhang Y.N., Sial M.U., Yu H.T., Cao A.C. Replacing methyl bromide with a combination of 1,3-dichloropropene and

metam sodium for cucumber production in China. PLoS ONE. 2017;12:e0188137. doi: 10.1371/journal.pone.0188137.
382. Rajashekar Y., Bakthavatsalam N., Shivanandappa T. Botanicals as grain protectants. Psyche. 2012;2012:646740. doi: 10.1155/2012/646740.
383. Ammann K. Integrated farming: Why organic farmers should use transgenic crops. New Biotechnol. 2008;25:101–107. doi: 10.1016/j.nbt.2008.08.012
384. Thamsborg S.M., Roepstorff A., Larsen M. Integrated and biological control of parasites in organic and conventional production systems. Vet. Parasitol. 1999;84:169–186. doi: 10.1016/S0304-4017(99)00035-7.
385. Arunrat N., Sereenonchai S., Wang C. Carbon footprint and predicting the impact of climate change on carbon sequestration ecosystem services of organic rice farming and conventional rice farming: A case study in Phichit province, Thailand. J. Environ. Manag. 2021;289:112458. doi: 10.1016/j.jenvman.2021.112458.
386. Hernández-Suárez E.M., Beitia F. Sustainable management methods of orchard insect pests. Insects. 2021;12:80. doi: 10.3390/insects12010080.
387. Katan J. Physical and cultural methods for the management of soil-borne pathogens. Crop Prot. 2000;19:725–731. doi: 10.1016/S0261-2194(00)00096-X.
388. Conti S., Villari G., Faugno S., Melchionna G., Somma S., Caruso G. Effects of organic vs. conventional farming system on yield and quality of strawberry grown as an annual or biennial crop in southern Italy. Sci. Hortic. 2014;180:63–71. doi: 10.1016/j.scienta.2014.10.015.
389. Villaverde J.J., Sevilla-Morán B., Sandín-España P., López-Goti C., Alonso-Prados J.L. Challenges of biopesticides under the European Regulation (EC) No. 1107/2009: An overview of new trends in residue analysis. In: Atta-ur-

Rahman, editor. Studies in Natural Products Chemistry. Elsevier; Oxford, UK: 2014. pp. 437–482.

390. Smolińska U., Kowalska B. Biological control of the soil-borne fungal pathogen Sclerotinia sclerotiorum—A review. J. Plant Pathol. 2018;100:1–12. doi: 10.1007/s42161-018-0023-0.

391. Schmitt A., Seddon B. Biocontrol of plant pathogens with microbial BCAs and plant extracts—Advantages and disadvantages of single and combined use; Proceedings of the 14th International Reinhardsbrunn Symposium on Modern Fungicides and Antifungal Compounds; Friedrichroda, Germany. 25–29 April 2004.

392. Barzman M., Bàrberi P., Birch A.N.E., Boonekamp P., Dachbrodt-Saaydeh S., Graf B., Hommel B., Jensen J.E., Kiss J., Kudsk P., et al. Eight principles of integrated pest management. Agron. Sustain. Dev. 2015;35:1199–1215. doi: 10.1007/s13593-015-0327-9.

393. Deguine J.P., Aubertot J.N., Flor R.J., Lescourret F., Wyckhuys K.A.G., Ratnadass A. Integrated pest management: Good intentions, hard realities. A review. Agron. Sustain. Dev. 2021;41:38. doi: 10.1007/s13593-021-00689-w.

394. Kabir M.H., Rainis R. Adoption and intensity of integrated pest management (IPM) vegetable farming in Bangladesh: An approach to sustainable agricultural development. Environ. Dev. Sustain. 2015;17:1413–1429. doi: 10.1007/s10668-014-9613-y

395. Fielding B.C., Knowles C.L., Vries F.A., Klaasen J.A. Testing of eight medicinal plant extracts in combination with kresoxim-methyl for integrated control of Botrytis cinerea in apples. Agriculture. 2015;5:400–411. doi: 10.3390/agriculture5030400.

396. Shlevin E., Gamliel A., Katan J., Shtienberg D. Multi-study analysis of the added benefits of combining soil solarization with fumigants or non-chemical measures. Crop Prot. 2018;111:58–65. doi: 10.1016/j.cropro.2018.05.001.

397. Ait Elallem B., Sobeh M., Boularbah A., Yasri A. Chemically degraded soil rehabilitation process using medicinal and aromatic plants: Review. Environ. Sci. Pollut. Res. 2021;28:73–93. doi: 10.1007/s11356-020-10742-y.

398. Ebadollahi A., Ziaee M., Palla F. Essential oils extracted from different species of the Lamiaceae plant family as prospective bioagents against several detrimental pests. Molecules. 2020;25:1556. doi: 10.3390/molecules25071556.

399. Posgay M., Greff B., Kapcsándi V., Lakatos E. Effect of Thymus vulgaris L. essential oil and thymol on the microbiological properties of meat and meat products: A review. Heliyon. 2022;8:e10812. doi: 10.1016/j.heliyon.2022.e10812.

400. Song B.Z., Han Z.H. Assessment of the biocontrol effects of three aromatic plants on multiple trophic levels of the arthropod community in an agroforestry ecosystem. Ecol. Entomol. 2020;45:831–839. doi: 10.1111/een.12858.

401. Ehlers B.K., Berg M.P., Staudt M., Holmstrup M., Glasius M., Ellers J., Tomiolo S., Madsen R.B., Slotsbo S., Penuelas J. Plant secondary compounds in soil and their role in belowground species interactions. Trends Ecol. Evol. 2020;35:716–730. doi: 10.1016/j.tree.2020.04.001.

402. Thakur M., Bhattacharya S., Khosla P.K., Puri S. Improving production of plant secondary metabolites through biotic and abiotic elicitation. J. Appl.

Res. Med. Aromat. Plants. 2019;12:1–12. doi: 10.1016/j.jarmap.2018.11.004.

403. Bais H.P., Walker T.S., Schweizer H.P., Vivanco J.M. Root specific elicitation and antimicrobial activity of rosmarinic acid in hairy root cultures of Ocimum basilicum. Plant Physiol. Biochem. 2002;40:983–995. doi: 10.1016/S0981-9428(02)01460-2.

404. Colpas F.T., Schwan-Estrada K.R.F., Stangarlin J.R., Ferrarese M.D.L., Scapim C.A., Bonaldo S.M. Induction of plant defense responses by Ocimum gratissimum L. (Lamiaceae) leaf extracts. Summa Phytopathol. 2009;35:191–195. doi: 10.1590/S0100-54052009000300005.

405. Zhang Y., Han M.Z., Song M.N., Tian J., Song B.Z., Hu Y.J., Zhang J., Yao Y.C. Intercropping with aromatic plants increased the soil organic matter content and changed the microbial community in a pear orchard. Front. Microbiol. 2021;12:616932. doi: 10.3389/fmicb.2021.616932.

406. Borges D.F., Lopes E.A., Moraes A.R.F., Soares M.S., Visôtto L.E., Oliveira C.R., Valente V.M.M. Formulation of botanicals for the control of plant-pathogens: A review. Crop Prot. 2018;110:135–140. doi: 10.1016/j.cropro.2018.04.003.

407. Mandal S.M., Chakraborty D., Dey S. Phenolic acids act as signaling molecules in plant–microbe symbioses. Plant Signal. Behav. 2010;5:359–368. doi: 10.4161/psb.5.4.10871.

408. Misra P., Maji D., Awasthi A., Pandey S.S., Yadav A., Pandey A., Saikia D., Babu C.S.V., Kalra A. Vulnerability of soil microbiome to monocropping of medicinal and aromatic plants and its restoration through intercropping and organic amendments. Front. Microbiol. 2019;10:2604. doi: 10.3389/fmicb.2019.02604.

409. Mohammadi H., Dashi R., Farzaneh M., Parviz L., Hashempour H. Effects of beneficial root pseudomonas on morphological, physiological, and phytochemical characteristics of Satureja hortensis (Lamiaceae) under water stress. Braz. J. Bot. 2017;40:41–48. doi: 10.1007/s40415-016-0319-2.

410. Sharifi-Rad J., Sureda A., Tenore G.C., Daglia M., Sharifi-Rad M., Valussi M., Tundis R., Sharifi-Rad M., Loizzo M.R., Ademiluyi A.O., et al. Biological activities of essential oils: From plant chemoecology to traditional healing systems. Molecules. 2017;22:70. doi: 10.3390/molecules22010070.

411. Wenke K., Kai M., Piechulla B. Belowground volatiles facilitate interactions between plant roots and soil organisms. Planta. 2010;231:499–506. doi: 10.1007/s00425-009-1076-2.

412. Parolin P., Bresch C., Poncet C., Desneux N. Introducing the term 'Biocontrol Plants' for integrated pest management. Sci. Agric. 2014;71:77–80. doi: 10.1590/S0103-90162014000100011

413. Saha A., Basak B.B. Scope of value addition and utilization of residual biomass from medicinal and aromatic plants. Ind. Crops Prod. 2020;145:111979. doi: 10.1016/j.indcrop.2019.111979.

414. Zaccardelli M., Sorrentino R., Caputo M., Scotti R., De Falco E., Pane C. Stepwise-selected Bacillus amyloliquefaciens and B. subtilis strains from composted aromatic plant waste able to control soil-borne diseases. Agriculture. 2020;10:30. doi: 10.3390/agriculture10020030

415. Regnault-Roger C., Philogène B.J.R. Past and current prospects for the use of botanicals and plant allelochemicals in integrated pest management. Pharm. Biol. 2008;46:41–52. doi: 10.1080/13880200701729794.

416. Jacobson M. Botanical pesticides. Past, present, and future. In: Arnason J.T., Philogène B.J.R., Morand P., editors. Insecticides of Plant Origin. American Chemical Society; Washington, DC, USA: 1989. [(accessed on 17 January 2023)]. pp. 1–10. Available online: https://pubs.acs.org/doi/pdf/10.1021/bk-1989-0387.ch001.

417. Brahmachari G. Neem—An omnipotent plant: A retrospection. ChemBioChem. 2004;5:408–421. doi: 10.1002/cbic.200300749.

418. Guo Y.Y., Chen J.B., Ren D., Du B., Wu L., Zhang Y.Y., Wang Z.Y., Qian S. Synthesis of osthol-based botanical fungicides and their antifungal application in crop protection. Bioorg. Med. Chem. 2021;40:116184. doi: 10.1016/j.bmc.2021.116184.

419. Mulugeta T., Muhinyuza J.B., Gouws-Meyer R., Matsaunyane L., Andreasson E., Alexandersson E. Botanicals and plant strengtheners for potato and tomato cultivation in Africa. J. Integr. Agric. 2020;19:406–427. doi: 10.1016/S2095-3119(19)62703-6.

420. Asbahani A.E., Miladi K., Badri W., Sala M., Addi E.H.A., Casabianca H., Mousadik A.E., Hartmann D., Jilale A., Renaud F.N.R., et al. Essential oils: From extraction to encapsulation. Int. J. Pharm. 2015;483:220–243. doi: 10.1016/j.ijpharm.2014.12.069.

421. Fierascu R.C., Fierascu I., Baroi A.M., Ortan A. Selected aspects related to medicinal and aromatic plants as alternative sources of bioactive compounds. Int. J. Mol. Sci. 2021;22:1521. doi: 10.3390/ijms22041521.

422. Ribeiro-Santos R., Carvalho-Costa D., Cavaleiro C., Costa H.S., Albuquerque T.G., Castilho M.C., Ramos F., Melo N.R., Sanches-Silva A. A novel insight on an ancient aromatic plant: The rosemary (Rosmarinus

officinalis L.) Trends Food Sci. Technol. 2015;45:355–368. doi: 10.1016/j.tifs.2015.07.015.
423. Guesmi F., Saidi I., Bouzenna H., Hfaiedh N., Landoulsi A. Phytocompound variability, antioxidant and antibacterial activities, anatomical features of glandular and aglandular hairs of Thymus hirtus Willd. Ssp. algeriensis Boiss. and Reut. over developmental stages. S. Afr. J. Bot. 2019;127:234–243. doi: 10.1016/j.sajb.2019.09.004.
424. Kundu A., Dutta A., Mandal A., Negi L., Malik M., Puramchatwad R., Antil J., Singh A., Rao U., Saha S., et al. A comprehensive in vitro and in silico analysis of nematicidal action of essential oils. Front. Plant Sci. 2021;11:614143. doi: 10.3389/fpls.2020.614143.
425. Nazzaro F., Fratianni F., De Martino L., Coppola R., De Feo V. Effect of essential oils on pathogenic bacteria. Pharmaceuticals. 2013;6:1451–1474. doi: 10.3390/ph6121451.
426. Mota I., Sánchez-Sánchez J., Pedro L.G., Sousa M.J. Composition variation of the essential oil from Ocimum basilicum L. cv. Genovese Gigante in response to Glomus intraradices and mild water stress at different stages of growth. Biochem. Syst. Ecol. 2020;90:104021. doi: 10.1016/j.bse.2020.104021.
427. Schulz-Bohm K., Gerards S., Hundscheid M., Melenhorst J., de Boer W., Garbeva P. Calling from distance: Attraction of soil bacteria by plant root volatiles. ISME J. 2018;12:1252–1262. doi: 10.1038/s41396-017-0035-3.
428. Gershenzon J., McConkey M.E., Croteau R.B. Regulation of monoterpene accumulation in leaves of peppermint. Plant Physiol. 2000;122:205–214. doi: 10.1104/pp.122.1.205

429. van Roon A., Parsons J.R., te Kloeze A.M., Govers H.A.J. Fate and transport of monoterpenes through soils. Part I. Prediction of temperature dependent soil fate model input-parameters. Chemosphere. 2005;61:599–609. doi: 10.1016/j.chemosphere.2005.02.081.
430. Song B.Z., Tang G.B., Sang X.S., Zhang J., Yao Y.C., Wiggins N. Intercropping with aromatic plants hindered the occurrence of Aphis citricola in an apple orchard system by shifting predator–prey abundances. Biocontrol Sci. Technol. 2013;23:381–395. doi: 10.1080/09583157.2013.763904.
431. Vokou D., Margaris N.S., Lynch J.M. Effects of volatile oils from aromatic shrubs on soil microorganisms. Soil Biol. Biochem. 1984;16:509–513. doi: 10.1016/0038-0717(84)90060-9.
432. Jouini A., Verdeguer M., Pinton S., Araniti F., Palazzolo E., Badalucco L., Laudicina V.A. Potential effects of essential oils extracted from Mediterranean aromatic plants on target weeds and soil microorganisms. Plants. 2020;9:1289. doi: 10.3390/plants9101289.
433. Benomari F.Z., Andreu V., Kotarba J., Dib M.E.A., Bertrand C., Muselli A., Costa J., Djabou N. Essential oils from Algerian species of Mentha as new bio-control agents against phytopathogen strains. Environ. Sci. Pollut. Res. 2018;25:29889–29900. doi: 10.1007/s11356-017-9991-4.
434. Soković M., Tzakou O., Pitarokili D., Couladis M. Antifungal activities of selected aromatic plants growing wild in Greece. Nahrung/Food. 2002;46:317–320. doi: 10.1002/1521-3803(20020901)46:5<317::AID-FOOD317>3.0.CO;2-B.

435. Kalemba D., Kunicka A. Antibacterial and antifungal properties of essential oils. Curr. Med. Chem. 2003;10:813–829. doi: 10.2174/0929867033457719.

436. Radaelli M., da Silva B.P., Weidlich L., Hoehne L., Flach A., da Costa L.A.M.A., Ethur E.M. Antimicrobial activities of six essential oils commonly used as condiments in Brazil against Clostridium perfringens. Braz. J. Microbiol. 2016;47:424–430. doi: 10.1016/j.bjm.2015.10.001.

437. Dhifi W., Bellili S., Jazi S., Bahloul N., Mnif W. Essential oils' chemical characterization and investigation of some biological activities: A critical review. Medicines. 2016;3:25. doi: 10.3390/medicines3040025.

438. Karpiński T.M. Essential oils of Lamiaceae family plants as antifungals. Biomolecules. 2020;10:103. doi: 10.3390/biom10010103.

439. Wogiatzi E., Gougoulias N., Papachatzis A., Vagelas I., Chouliaras N. Chemical composition and antimicrobial effects of Greek Origanum species essential oil. Biotechnol. Biotechnol. Equip. 2009;23:1322–1324. doi: 10.1080/13102818.2009.10817662.

440. Pitarokili D., Tzakou O., Kalamarakis A. Activity of the essential oil of Salvia pomifera L. ssp. calycina (Sm.) Hayek against soil borne pathogens. J. Essent. Oil Res. 2002;14:72–75. doi: 10.1080/10412905.2002.9699769.

441. Aguilar-Veloz L.M., Calderón-Santoyo M., Vázquez González Y., Ragazzo-Sánchez J.A. Application of essential oils and polyphenols as natural antimicrobial agents in postharvest treatments: Advances and challenges. Food Sci. Nutr. 2020;8:2555–2568. doi: 10.1002/fsn3.1437.

442. Nieto G. Biological activities of three essential oils of the Lamiaceae family. Medicines. 2017;4:63. doi: 10.3390/medicines4030063.

443. Upadhyay N., Singh V.K., Dwivedy A.K., Das S., Chaudhari A.K., Dubey N.K. Cistus ladanifer L. essential oil as a plant based preservative against molds infesting oil seeds, aflatoxin B1 secretion, oxidative deterioration and methylglyoxal biosynthesis. LWT. 2018;92:395–403. doi: 10.1016/j.lwt.2018.02.040.

444. Sempere-Ferre F., Asamar J., Castell V., Roselló J., Santamarina M.P. Evaluating the antifungal potential of botanical compounds to control Botryotinia fuckeliana and Rhizoctonia solani. Molecules. 2021;26:2472. doi: 10.3390/molecules26092472.

445. Marei G.I.K., Abdel Rasoul M.A., Abdelgaleil S.A.M. Comparative antifungal activities and biochemical effects of monoterpenes on plant pathogenic fungi. Pestic. Biochem. Physiol. 2012;103:56–61. doi: 10.1016/j.pestbp.2012.03.004.

446. Lee S.O., Choi G.J., Jang K.S., Lim H.K., Cho K.Y., Kim J.C. Antifungal activity of five plant essential oils as fumigant against postharvest and soilborne plant pathogenic fungi. Plant Pathol. J. 2007;23:97–102. doi: 10.5423/PPJ.2007.23.2.097.

447. Bi Y., Jiang H., Hausbeck M.K., Hao J.J. Inhibitory effects of essential oils for controlling Phytophthora capsici. Plant Dis. 2012;96:797–803. doi: 10.1094/PDIS-11-11-0933.

448. Laquale S., Candido V., Avato P., Argentieri M.P., D'Addabbo T. Essential oils as soil biofumigants for the control of the root-knot nematode Meloidogyne incognita on tomato. Ann. Appl. Biol. 2015;167:217–224. doi: 10.1111/aab.12221.

449. Greff B., Lakatos E., Szigeti J., Varga L. Co-composting with herbal wastes: Potential effects of essential oil residues on microbial pathogens during

composting. Crit. Rev. Environ. Sci. Technol. 2021;51:457–511. doi: 10.1080/10643389.2020.1732780.

450. Mabrouki H., Duarte C.M.M., Akretche D.E. Estimation of total phenolic contents and in vitro antioxidant and antimicrobial activities of various solvent extracts of Melissa officinalis L. Arab. J. Sci. Eng. 2018;43:3349–3357. doi: 10.1007/s13369-017-3000-6.

451. Proestos C., Sereli D., Komaitis M. Determination of phenolic compounds in aromatic plants by RP-HPLC and GC-MS. Food Chem. 2006;95:44–52. doi: 10.1016/j.foodchem.2004.12.016.

452. Costa D.C., Costa H.S., Albuquerque T.G., Ramos F., Castilho M.C., Sanches-Silva A. Advances in phenolic compounds analysis of aromatic plants and their potential applications. Trends Food Sci. Technol. 2015;45:336–354. doi: 10.1016/j.tifs.2015.06.009.

453. Seró R., Núñez N., Núñez O., Camprubí A., Grases J.M., Saurina J., Moyano E., Calvet C. Modified distribution in the polyphenolic profile of rosemary leaves induced by plant inoculation with an arbuscular mycorrhizal fungus. J. Sci. Food Agric. 2019;99:2966–2973. doi: 10.1002/jsfa.9510.

454. Durazzo A., Lucarini M., Souto E.B., Cicala C., Caiazzo E., Izzo A.A., Novellino E., Santini A. Polyphenols: A concise overview on the chemistry, occurrence, and human health. Phytother. Res. 2019;33:2221–2243. doi: 10.1002/ptr.6419.

455. Sakihama Y., Cohen M.F., Grace S.C., Yamasaki H. Plant phenolic antioxidant and prooxidant activities: Phenolics-induced oxidative damage mediated by metals in plants. Toxicology. 2002;177:67–80. doi: 10.1016/S0300-483X(02)00196-8.

456. D'Addabbo T., Laquale S., Lovelli S., Candido V., Avato P. Biocide plants as a sustainable tool for the control of pests and pathogens in vegetable cropping systems. Ital. J. Agron. 2014;9:137. doi: 10.4081/ija.2014.616.
457. Eghbaliferiz S., Iranshahi M. Prooxidant activity of polyphenols, flavonoids, anthocyanins and carotenoids: Updated review of mechanisms and catalyzing metals. Phytother. Res. 2016;30:1379–1391. doi: 10.1002/ptr.5643.
458. Sytar O., Hemmerich I., Zivcak M., Rauh C., Brestic M. Comparative analysis of bioactive phenolic compounds composition from 26 medicinal plants. Saudi J. Biol. Sci. 2018;25:631–641. doi: 10.1016/j.sjbs.2016.01.036.
459. Brown N., John J.A., Shahidi F. Polyphenol composition and antioxidant potential of mint leaves. Food Prod. Process. Nutr. 2019;1:1. doi: 10.1186/s43014-019-0001-8.
460. Duque-Soto C., Borrás-Linares I., Quirantes-Piné R., Falcó I., Sánchez G., Segura-Carretero A., Lozano-Sánchez J. Potential antioxidant and antiviral activities of hydroethanolic extracts of selected Lamiaceae species. Foods. 2022;11:1862. doi: 10.3390/foods11131862.
461. Mojzer E.B., Hrnčič M.K., Škerget M., Knez Ž., Bren U. Polyphenols: Extraction methods, antioxidative action, bioavailability and anticarcinogenic effects. Molecules. 2016;21:901. doi: 10.3390/molecules21070901.
462. Bouarab-Chibane L., Forquet V., Lantéri P., Clément Y., Léonard-Akkari L., Oulahal N., Degraeve P., Bordes C. Antibacterial properties of polyphenols: Characterization and QSAR (Quantitative Structure–Activity Relationship) models. Front. Microbiol. 2019;10:829. doi: 10.3389/fmicb.2019.00829.

463. Proestos C., Chorianopoulos N., Nychas G.J.E., Komaitis M. RP-HPLC analysis of the phenolic compounds of plant extracts. Investigation of their antioxidant capacity and antimicrobial activity. J. Agric. Food Chem. 2005;53:1190–1195. doi: 10.1021/jf040083t.

464. Cetin-Karaca H., Newman M.C. Antimicrobial efficacy of plant phenolic compounds against Salmonella and Escherichia coli. Food Biosci. 2015;11:8–16. doi: 10.1016/j.fbio.2015.03.002.

465. Daglia M. Polyphenols as antimicrobial agents. Curr. Opin. Biotechnol. 2012;23:174–181. doi: 10.1016/j.copbio.2011.08.007.

466. Shalaby S., Horwitz B.A. Plant phenolic compounds and oxidative stress: Integrated signals in fungal–plant interactions. Curr. Genet. 2015;61:347–357. doi: 10.1007/s00294-014-0458-6.

467. Gao X.Z., Xu Z.J., Liu G.T., Wu J. Polyphenols as a versatile component in tissue engineering. Acta Biomater. 2021;119:57–74. doi: 10.1016/j.actbio.2020.11.004.

468. Khameneh B., Iranshahy M., Soheili V., Bazzaz B.S.F. Review on plant antimicrobials: A mechanistic viewpoint. Antimicrob. Resist. Infect. Control. 2019;8:118. doi: 10.1186/s13756-019-0559-6.

469. Summers C.B., Felton G.W. Prooxidant effects of phenolic acids on the generalist herbivore Helicoverpa zea (Lepidoptera: Noctuidae): Potential mode of action for phenolic compounds in plant anti-herbivore chemistry. Insect Biochem. Mol. Biol. 1994;24:943–953. doi: 10.1016/0965-1748(94)90023-X.

470. Simmonds M.S.J. Flavonoid–insect interactions: Recent advances in our knowledge. Phytochemistry. 2003;64:21–30. doi: 10.1016/S0031-9422(03)00293-0.

471. Widmer T.L., Laurent N. Plant extracts containing caffeic acid and rosmarinic acid inhibit zoospore germination of Phytophthora spp. pathogenic to Theobroma cacao. Eur. J. Plant Pathol. 2006;115:377–388. doi: 10.1007/s10658-006-9024-5.

472. Al-Huqail A.A., Behiry S.I., Salem M.Z.M., Ali H.M., Siddiqui M.H., Salem A.Z.M. Antifungal, antibacterial, and antioxidant activities of Acacia saligna (Labill.) H. L. Wendl. flower extract: HPLC analysis of phenolic and flavonoid compounds. Molecules. 2019;24:700. doi: 10.3390/molecules24040700.

473. Matsubara Y., Yamashita Y., Liu J. Antifungal and antioxidative ability in Lamiaceae herbs. Acta Hortic. 2015;1105:109–114. doi: 10.17660/ActaHortic.2015.1105.16.

474. Ahmad H., Matsubara Y. Suppression of Fusarium wilt in Cyclamen by using sage water extract and identification of antifungal metabolites. Australas. Plant Pathol. 2020;49:213–220. doi: 10.1007/s13313-020-00689-6.

475. López V., Akerreta S., Casanova E., García-Mina J.M., Cavero R.Y., Calvo M.I. In vitro antioxidant and anti-rhizopus activities of Lamiaceae herbal extracts. Plant Foods Hum. Nutr. 2007;62:151–155. doi: 10.1007/s11130-007-0056-6.

476. Skrzypek E., Repka P., Stachurska-Swakoń A., Barabasz-Krasny B., Możdżeń K. Allelopathic effect of aqueous extracts from the leaves of peppermint (Mentha piperita L.) on selected physiological processes of common sunflower (Helianthus annuus L.) Not. Bot. Horti Agrobot. 2015;43:335–342. doi: 10.15835/nbha43210034.

477. Isman M.B. Botanical insecticides, deterrents, and repellents in modern agriculture and an increasingly regulated world. Annu. Rev. Entomol. 2006;51:45–66. doi: 10.1146/annurev.ento.51.110104.151146.

478. NÉBIH Fertilizer Database. [(accessed on 17 January 2023)]; Available online: https://termesnovelo.nebih.gov.hu/Engedelykereso/kereso.

479. NÉBIH Pesticide Database. [(accessed on 17 January 2023)]; Available online: https://novenyvedoszer.nebih.gov.hu/Engedelykereso/kereso.

480. EU Pesticide Database. [(accessed on 17 January 2023)]. Available online: https://food.ec.europa.eu/plants/pesticides/eu-pesticides-database_en.

481. Lin C., Owen S.M., Peñuelas J. Volatile organic compounds in the roots and rhizosphere of Pinus spp. Soil Biol. Biochem. 2007;39:951–960. doi: 10.1016/j.soilbio.2006.11.007.

482. Li X.G., de Boer W., Zhang Y.A., Ding C.F., Zhang T.L., Wang X.X. Suppression of soil-borne Fusarium pathogens of peanut by intercropping with the medicinal herb Atractylodes lancea. Soil Biol. Biochem. 2018;116:120–130. doi: 10.1016/j.soilbio.2017.09.029.

483. Mamolos A.P., Kalburtji K.L. Significance of allelopathy in crop rotation. J. Crop Prod. 2001;4:197–218. doi: 10.1300/J144v04n02_06.

484. Verma R.K., Yadav A., Rahman L.U., Kalra A., Patra D.D. Influence the status of soil chemical and biological properties by intercropping. Int. J. Recycl. Org. Waste Agric. 2014;3:46. doi: 10.1007/s40093-014-0046-2.

485. Vukicevich E., Lowery T., Bowen P., Úrbez-Torres J.R., Hart M. Cover crops to increase soil microbial diversity and mitigate decline in perennial agriculture. A review. Agron. Sustain. Dev. 2016;36:48. doi: 10.1007/s13593-016-0385-7.

486. Khan K., Tewari S.K., Semwal M., Verma R.K. Effects of aromatic crops based cropping systems and residue returning on soil CO2 emission compared to fallow soil. J. Appl. Res. Med. Aromat. Plants. 2022;26:100351. doi: 10.1016/j.jarmap.2021.100351.
487. Khan K., Verma R.K. Diversifying cropping systems with aromatic crops for better productivity and profitability in subtropical north Indian plains. Ind. Crops Prod. 2018;115:104–110. doi: 10.1016/j.indcrop.2018.02.004
488. Raveau R., Fontaine J., Hijri M., Sahraoui A.L.H. The aromatic plant clary sage shaped bacterial communities in the roots and in the trace element-contaminated soil more than mycorrhizal inoculation—A two-year monitoring field trial. Front. Microbiol. 2020;11:586050. doi: 10.3389/fmicb.2020.586050.
489. Zhou H.L., Liu X.Y., Zhou C., Han S.F., Xu F.R., Dong X. Root-associated microbiomes of Panax notoginseng under the combined effect of plant development and Alpinia officinarum Hance essential oil. Molecules. 2022;27:6014. doi: 10.3390/molecules27186014.
490. Bafana A. Diversity and metabolic potential of culturable root-associated bacteria from Origanum vulgare in sub-Himalayan region. World J. Microbiol. Biotechnol. 2013;29:63–74. doi: 10.1007/s11274-012-1158-3.
491. Karthikeyan B., Jaleel C.A., Lakshmanan G.M.A., Deiveekasundaram M. Studies on rhizosphere microbial diversity of some commercially important medicinal plants. Colloids Surf. B Biointerfaces. 2008;62:143–145. doi: 10.1016/j.colsurfb.2007.09.004.
492. Adamović D., Đalović I., Mrkovački N. Microbial abundance in rhizosphere of medicinal and aromatic plant species in conventional and organic growing systems. Ratar. Povrt. 2015;52:1–6. doi: 10.5937/ratpov51-7177.

493. Jisha C.K., Bauddh K., Shukla S.K. Phytoremediation and bioenergy production efficiency of medicinal and aromatic plants. In: Bauddh K., Singh B., Korstad J., editors. Phytoremediation Potential of Bioenergy Plants. Springer; Singapore: 2017. pp. 287–304.

494. Chen L.J., Wu X.Q., Xu Y., Wang B.L., Liu S., Niu J.F., Wang Z. Microbial diversity and community structure changes in the rhizosphere soils of Atractylodes lancea from different planting years. Plant Signal. Behav. 2021;16:1854507. doi: 10.1080/15592324.2020.1854507.

495. Karkanis A., Alexiou A., Katsaros C., Petropoulos S. Allelopathic activity of spearmint (Mentha spicata L.) and peppermint (Mentha × piperita L.) reduces yield, growth, and photosynthetic rate in a succeeding crop of maize (Zea mays L.) Agronomy. 2019;9:461. doi: 10.3390/agronomy9080461.

496. Dittrich F., Iserloh T., Treseler C.H., Hüppi R., Ogan S., Seeger M., Thiele-Bruhn S. Crop diversification in viticulture with aromatic plants: Effects of intercropping on grapevine productivity in a steep-slope vineyard in the Mosel area, Germany. Agriculture. 2021;11:95. doi: 10.3390/agriculture11020095.

497. Chand S., Singh D., Madegowda M., Sikka A. Medicinal and aromatic plants for soil and water conservation in Nilgiris, Tamil Nadu, India: An economic analysis. Mol. Soil Biol. 2015;665:1–6. doi: 10.5376/msb.2015.06.0001.

498. Dhima K.V., Vasilakoglou I.B., Gatsis T.D., Panou-Philotheou E., Eleftherohorinos I.G. Effects of aromatic plants incorporated as green manure on weed and maize development. Field Crops Res. 2009;110:235–241. doi: 10.1016/j.fcr.2008.09.005.

499. Kamil D., Kumar R., Sinha A. Effect of green manuring of Crotolaria juncea L. on some soil-borne pathogens. Indian Phytopathol. 2009;62:304–309.
500. Dojima T., Craker L.E. Potential benefits of soil microorganisms on medicinal and aromatic plants. In: Jeliazkov V.D., Cantrell C.L., editors. Medicinal and Aromatic Crops: Production, Phytochemistry, and Utilization. American Chemical Society; Washington, DC, USA: 2016. pp. 75–90.
501. Kushwaha R.K., Rodrigues V., Kumar V., Patel H., Raina M., Kumar D. Soil microbes–medicinal plants interactions: Ecological diversity and future prospect. In: Varma A., Tripathi S., Prasad R., editors. Plant Microbe Symbiosis. Springer; Cham, Switzerland: 2020. pp. 263–286.
502. Castronovo L.M., Calonico C., Ascrizzi R., Del Duca S., Delfino V., Chioccioli S., Vassallo A., Strozza I., De Leo M., Biffi S., et al. The cultivable bacterial microbiota associated to the medicinal plant Origanum vulgare L.: From antibiotic resistance to growth-inhibitory properties. Front. Microbiol. 2020;11:862. doi: 10.3389/fmicb.2020.00862.
503. Tiwari S., Pandey R., Gross A. Identification of rhizospheric microorganisms that manages root knot nematode and improve oil yield in sweet basil (Ocimum basilicum L.) Agronomy. 2021;11:570. doi: 10.3390/agronomy11030570.
504. Chowdhary K., Kaushik N. Biodiversity study and potential of fungal endophytes of peppermint and effect of their extract on chickpea rot pathogens. Arch. Phytopathol. Plant Prot. 2018;51:139–155. doi: 10.1080/03235408.2018.1440707.
505. Rao E.V.S.P. Medicinal and aromatic plants for crop diversification and their agronomic implications. Indian J. Agron. 2009;54:215–220.

506. Ahmad H., Kobayashi M., Matsubara Y. Changes in secondary metabolites and free amino acid content in tomato with Lamiaceae herbs companion planting. Am. J. Plant Sci. 2020;11:1878–1889. doi: 10.4236/ajps.2020.1112134.
507. Parolin P., Bresch C., Poncet C., Suay-Cortez R., Van Oudenhove L. Testing basil as banker plant in IPM greenhouse tomato crops. Int. J. Pest Manag. 2015;61:235–242. doi: 10.1080/09670874.2015.1042414.
508. Carvalho L.M., Nunes M.U.C., Oliveira I.R., Leal M.L.S. Produtividade do tomateiro em cultivo solteiro e consorciado com espécies aromáticas e medicinais. Hortic. Bras. 2009;27:458–464. doi: 10.1590/S0102-05362009000400010.
509. Puig C.G., Álvarez-Iglesias L., Reigosa M.J., Pedrol N. Eucalyptus globulus leaves incorporated as green manure for weed control in maize. Weed Sci. 2013;61:154–161. doi: 10.1614/WS-D-12-00056.1.
510. Thakur M., Kumar R. Microclimatic buffering on medicinal and aromatic plants: A review. Ind. Crops Prod. 2021;160:113144. doi: 10.1016/j.indcrop.2020.113144.
511. de Elguea-Culebras G.O., Bravo E.M., Sánchez-Vioque R. Potential sources and methodologies for the recovery of phenolic compounds from distillation residues of Mediterranean aromatic plants. An approach to the valuation of by-products of the essential oil market—A review. Ind. Crops Prod. 2022;175:114261. doi: 10.1016/j.indcrop.2021.114261.
512. Skendi A., Irakli M., Chatzopoulou P., Bouloumpasi E., Biliaderis C.G. Phenolic extracts from solid wastes of the aromatic plant essential oil industry: Potential uses in food applications. Food Chem. Adv. 2022;1:100065. doi: 10.1016/j.focha.2022.100065.

513. Basak B.B. Recycling of waste biomass and mineral powder for preparation of potassium-enriched compost. J. Mater. Cycles Waste Manag. 2018;20:1409–1415. doi: 10.1007/s10163-018-0699-4.
514. Yohalem D., Passey T. Amendment of soils with fresh and post-extraction lavender (Lavandula angustifolia) and lavandin (Lavandula × intermedia) reduce inoculum of Verticillium dahliae and inhibit wilt in strawberry. Appl. Soil Ecol. 2011;49:187–196. doi: 10.1016/j.apsoil.2011.05.006.
515. Kadoglidou K., Chalkos D., Karamanoli K., Eleftherohorinos I.G., Constantinidou H.I.A., Vokou D. Aromatic plants as soil amendments: Effects of spearmint and sage on soil properties, growth and physiology of tomato seedlings. Sci. Hortic. 2014;179:25–35. doi: 10.1016/j.scienta.2014.09.009.
516. Klein E., Katan J., Gamliel A. Combining residues of herb crops with soil heating for control of soilborne pathogens in a controlled laboratory system. Crop Prot. 2011;30:368–374. doi: 10.1016/j.cropro.2010.12.005.
517. Vokou D., Chalkos D., Karamanlidou G., Yiangou M. Activation of soil respiration and shift of the microbial population balance in soil as a response to Lavandula stoechas essential oil. J. Chem. Ecol. 2002;28:755–768. doi: 10.1023/A:1015236709767.
518. Gravanis F., Chouliaras N., Vagelas I., Gougoulias N. The effect of oregano (Origanum vulgare) as an alternative soil-borne pathogen control agent, on soil organic matter biodegradation and other soil chemical properties; Proceedings of the BCPC International Congress on Crop Science & Technology; Glasgow, UK. 31 October–2 November 2005.
519. Chouliaras N., Gravanis F., Vasilakoglou I., Gougoulias N., Vagelas I., Kapotis T., Wogiatzi E. The effect of basil (Ocimum basilicum L.) on soil

organic matter biodegradation and other soil chemical properties. J. Sci. Food Agric. 2007;87:2416–2419. doi: 10.1002/jsfa.2907.

520. Ainalidou A., Bouzoukla F., Menkissoglu-Spiroudi U., Vokou D., Karamanoli K. Impacts of decaying aromatic plants on the soil microbial community and on tomato seedling growth and metabolism: Suppression or stimulation? Plants. 2021;10:1848. doi: 10.3390/plants10091848.

521. Argyropoulou M.D., Karmezi M., Tsiafouli M., Chalkos D., Bountla A., Vokou D. Soil amendments with spearmint, peppermint and rosemary enhance the community of free-living nematodes and improve soil quality, while having strikingly different effects on plant growth. Life. 2022;12:1121. doi: 10.3390/life12081121.

522. Deka H., Deka S., Baruah C.K., Das J., Hoque S., Sarma H., Sarma N.S. Vermicomposting potentiality of Perionyx excavatus for recycling of waste biomass of java citronella—An aromatic oil yielding plant. Bioresour. Technol. 2011;102:11212–11217. doi: 10.1016/j.biortech.2011.09.102.

523. El-Helw N.O., El-Gendy A.O., El-Gebaly E., Hassan H.M., Rateb M.E., El-Nesr K.A. Characterization of natural bioactive compounds produced by isolated bacteria from compost of aromatic plants. J. Appl. Microbiol. 2019;126:443–451. doi: 10.1111/jam.14085.

524. Pane C., Piccolo A., Spaccini R., Celano G., Villecco D., Zaccardelli M. Agricultural waste-based composts exhibiting suppressivity to diseases caused by the phytopathogenic soil-borne fungi Rhizoctonia solani and Sclerotinia minor. Appl. Soil Ecol. 2013;65:43–51. doi: 10.1016/j.apsoil.2013.01.002.

525. Pandey R., Kalra A. Inhibitory effects of vermicompost produced from agro-waste of medicinal and aromatic plants on egg hatching in Meloidogyne incognita (Kofoid and White) Chitwood. Curr. Sci. 2010;98:833–835.
526. Singh R., Soni S.K., Awasthi A., Kalra A. Evaluation of vermicompost doses for management of root-rot disease complex in Coleus forskohlii under organic field conditions. Australas. Plant Pathol. 2012;41:397–403. doi: 10.1007/s13313-012-0134-6.
527. Zhou Y., Selvam A., Wong J.W.C. Effect of Chinese medicinal herbal residues on microbial community succession and anti-pathogenic properties during co-composting with food waste. Bioresour. Technol. 2016;217:190–199. doi: 10.1016/j.biortech.2016.03.080.
528. Bajer T., Šilha D., Ventura K., Bajerová P. Composition and antimicrobial activity of the essential oil, distilled aromatic water and herbal infusion from Epilobium parviflorum Schreb. Ind. Crops Prod. 2017;100:95–105. doi: 10.1016/j.indcrop.2017.02.016.
529. Ozturk I., Tornuk F., Caliskan-Aydogan O., Durak M.Z., Sagdic O. Decontamination of iceberg lettuce by some plant hydrosols. LWT. 2016;74:48–54. doi: 10.1016/j.lwt.2016.06.067.
530. Tabti L., Dib M.E.A., Djabou N., Benyelles N.G., Paolini J., Costa J., Muselli A. Control of fungal pathogens of Citrus sinensis L. by essential oil and hydrosol of Thymus capitatus L. J. Appl. Bot. Food Qual. 2014;87:279–285. doi: 10.5073/JABFQ.2014.087.039.
531. Gaspar-Pintiliescu A., Mihai E., Stanciuc A.M., Anton R.E., Popescu A.F., Craciunescu O. Polyphenolic content, antioxidant and antimicrobial activity of hydrosols from some aromatic Lamiaceae plants. Proceedings. 2020;57:21. doi: 10.3390/proceedings2020057021.

532. Andrés M.F., González-Coloma A., Muñoz R., De la Peña F., Julio L.F., Burillo J. Nematicidal potential of hydrolates from the semi industrial vapor-pressure extraction of Spanish aromatic plants. Environ. Sci. Pollut. Res. 2018;25:29834–29840. doi: 10.1007/s11356-017-9429-z.

533. Ntalli N., Parlapani A.B., Tzani K., Samara M., Boutsis G., Dimou M., Menkissoglu-Spiroudi U., Monokrousos N. Thymus citriodorus (Schreb) botanical products as ecofriendly nematicides with bio-fertilizing properties. Plants. 2020;9:202. doi: 10.3390/plants9020202.

534. Sainz P., Andrés M.F., Martínez-Díaz R.A., Bailén M., Navarro-Rocha J., Díaz C.E., González-Coloma A. Chemical composition and biological activities of Artemisia pedemontana subsp. assoana essential oils and hydrolate. Biomolecules. 2019;9:558. doi: 10.3390/biom9100558.

535. Pino-Otín R., Val J., Ballestero D., Navarro E., Sánchez E., Mainar A.M. Impact of Artemisia absinthium hydrolate extracts with nematicidal activity on non-target soil organisms of different trophic levels. Ecotoxicol. Environ. Saf. 2019;180:565–574. doi: 10.1016/j.ecoenv.2019.05.055.

536. La Bella, S.; Virga, G.; Iacuzzi, N.; Licata, M.; Sabatino, L.; Consentino, B.B.; Leto, C.; Tuttolomondo, T. Effects of irrigation, peat-alternative substrate and plant habitus on the morphological and production characteristics of Sicilian rosemary (*Rosmarinus officinalis* L.) biotypes grown in pot. *Agriculture* **2021**, *11*, 13.

537. Pinto-Morales, F.; Retamal-Salgado, J.; Lopéz, M.D.; Zapata, N.; Vergara-Retamales, R.; Pinto-Poblete, A. The use of compost increases bioactive compounds and fruit. *Agriculture* **2022**, *12*, 98.

538. Angelini, L.G.; Clemente, C.; Tavarini, S. Pre-germination treatments, temperature, and light conditions improved seed germination of *Passiflora incarnata* L. *Agriculture* **2021**, *11*, 937.

539. Fernández-Sestelo, M.; Carrillo, J.M. Environmental effects on yield and composition of essential oil in wild populations of spike lavender (*Lavandula latifolia* Medik.). *Agriculture* **2020**, *10*, 626.

540. Lazzara, S.; Carrubba, A.; Napoli, E. Cultivating for the industry: Cropping experiences with *Hypericum perforatum* L. in a Mediterranean environment. *Agriculture* **2021**, *11*, 446.

541. Clemente, C.; Angelini, L.G.; Ascrizzi, R.; Tavarini, S. *Stevia rebaudiana* (Bertoni) as a multifunctional and sustainable crop for the Mediterranean climate. *Agriculture* **2021**, *11*, 123.

542. La Bella, S.; Rossini, F.; Licata, M.; Virga, G.; Ruggeri, R.; Iacuzzi, N.; Leto, C.; Tuttolomondo, T. Four-year study on the bio-agronomic response of biotypes of *Capparis spinosa* L. on the island of Linosa (Italy). *Agriculture* **2021**, *11*, 327.

543. Izardi, Z.; Rezaei Nejad, A.; Abadía, J. Iron chelate improves rooting in indole-3-butyric acid-treated rosemary (*Rosmarinus officinalis*) stem cuttings. *Agriculture* **2022**, *12*, 210.

544. Yang, X.; Li, Y.; Li, C.; Li, Q.; Qiao, B.; Shi, S.; Zhao, C. Enhancement of interplanting of *Ficus carica* L. with *Taxus cuspidata* Sieb. et Zucc. on growth of two plants. *Agriculture* **2021**, *11*, 1276.

545. Singh, V.S.; Dhawale, S.C.; Shakeel, F.; Faiyazuddin, M.; Alshehri, A. Antiarthritic potential of *Calotropis procera* leaf fractions in FCA-induced arthritic rats: Involvement of cellular inflammatory mediators and other

biomarkers yield in calafate grown in the central south of Chile. *Agriculture* **2021**, *11*, 68.

546. Barreca, S.; La Bella, S.; Maggio, A.; Licata, M.; Buscemi, S.; Leto, C.; Pace, A.; Tuttolomondo, T. Flavouring extra-virgin olive oil with aromatic and medicinal plants essential oils stabilizes oleic acid composition during photo-oxidative stress. *Agriculture* **2021**, *11*, 266.
547. Catania, P.; Gaglio, R.; Orlando, S.; Settanni, L.; Vallone, M. Design and implementation of a smart system to control aromatic herb dehydration process. *Agriculture* **2020**, *10*, 332.
548. Rossini, F.; Virga, G.; Loreti, P.; Iacuzzi, N.; Ruggeri, R.; Provenzano, M.E. Hops (*Humulus lupulus* L.) as a novel multipurpose crop for the Mediterranean region of Europe: Challenges and opportunities of their cultivation. *Agriculture* **2020**, *11*, 484.
549. Terpó, A. (1992). Taxonomy of Medicinal Plants. In: Hornok, L., ed.: Cultivation and Processing of Medicinal Plants. Publ. Akadémiai Kiadó, Budapest, pp. 42-55. [One of the few teaching materials on medicinal and aromatic plant production in English. Relatively comprehensive and contains most temperate belt species, as well as their detailed description – beginning with the botanical aspects through cultivation, processing and utilization].
550. Trease, E. and Evans, W. (2002). Pharmacognosy. In: W. Evans, ed.: W.B. Saunders, Edinburgh, UK, 15nd ed., 585p. [Well known and widely used textbook on Pharmacognosy. An absolute benchmark of its kind].
551. Zieger, E. and Taiz, L. (2006). Plant Physiology. Sinauer Associates Inc., Sunderland, U.S.A., 764p. [Standard textbook on plant physiology used by many universities, world wide].

552. Hornok, L., ed. (1992). *Cultivation and Processing of Medicinal Plants*. Publ. Akadémiai Kiadó, Budapest. [One of the few teaching materials on medicinal and aromatic plant production, in English. Relatively comprehensive. Deals mainly with temperate species, their detailed description, botanical aspects, cultivation, processing and utilization].

553. Tétényi, P. (1992). *Heredity and Variability of Medicinal Plants*. In: Hornok, L., ed.: Cultivation and Processing of Medicinal Plants. Publ. Akadémiai Kiadó, Budapest. p. 10-18. [A chapter by an outstanding MAPs breeder and chemo-taxonomist in one of the few teaching materials on medicinal and aromatic plant production, in English. Relatively comprehensive. Contains mostly temperate species, as well as their detailed description – beginning with the botanical aspects through cultivation, processing and utilization]

About Authors

Dr. Salah Abbas Zaidan Al-Tamimi

Ministry of Education, Iraq

Dr. Noor Sabah Nije Al Salihi

Ministry of Education, Iraq

FSC
www.fsc.org
MIX
Papier | Fördert gute Waldnutzung
FSC® C083411

Kolibri 360 GmbH
Ferdinand-Jühlke-Straße 7
99095 Erfurt, Deutschland
produktsicherheit@kolibri360.de

Druck:
CPI Druckdienstleistungen GmbH
im Auftrag der
Kolibri 360 GmbH
Ferdinand-Jühlke-Str. 7
99095 Erfurt